D1286298

ISLANDS

of

ABANDONMENT

Also by Cal Flyn

Thicker Than Water

ISLANDS

of

ABANDONMENT

Nature Rebounding in the Post-Human Landscape

Cal Flyn

VIKING

333.73 Fly

VIKING
An imprint of Penguin Random House LLC
penguinrandomhouse.com

First published in hardcover in Great Britain by William Collins, an imprint
of HarperCollins Publishers, London, in 2021.

First American edition published by Viking.

Grateful acknowledgment is made for permission
to reprint the following images:

Page 14 The Five Sisters bing. Copyright © 1975 by Dave Henniker.

Page 37 Copyright © Royal Commission on the Ancient
and Historical Monuments of Scotland.

Page 74 produced by Dr. Glenn Matlock, Associate Professor, Environmental
and Plant Biology Department, Ohio University, based on data from: Greeley,
William B, 1925, "The Relation of Geography to Timber Supply," *Economic
Geography* 1: 1–11; Williams, Michael, 1989, *Americans and Their Forests*,
Cambridge University Press, pp. 436–437; "Proportion of Land that is
Forested," US Forest Service Forest Inventory Analysis, 2007.

LIBRARY OF CONGRESS CATALOGING-IN-PUBLICATION DATA
Names: Flyn, Cal, author.
Title: Islands of abandonment : nature rebounding in
the post-human landscape / Cal Flyn.
Description: New York : Viking, [2021] | Includes
bibliographical references and index.
Identifiers: LCCN 2020041722 (print) | LCCN 2020041723 (ebook) |
ISBN 9781984878199 (hardcover) | ISBN 9781984878205 (ebook)
Subjects: LCSH: Restoration ecology. | Resilience (Ecology) | Environmental
disasters. | Ecosystem health. | Landscape changes. |
Nature—Effect of human beings on.
Classification: LCC QH541.15.R45 F59 2021 (print) | LCC QH541.15.R45
(ebook) | DDC 333.73/153—dc23
LC record available at https://lccn.loc.gov/2020041722
LC ebook record available at https://lccn.loc.gov/2020041723

Printed in the United States of America
1 3 5 7 9 10 8 6 4 2

Designed by Alexis Farabaugh

For Rich,

Who makes me so very happy

Contents

ISLANDS

of

ABANDONMENT

INVOCATION

Forth Islands, Scotland

I t is cool in the tunnels, not cold as it was outside. And dark, very dark. The air is almost still, but not quite—there's a trace of movement that scuffs the leaves that lie in low drifts along the edges where wall meets floor. Perhaps it's this that gives me the unnerving impression of not being entirely alone.

To reach the inner sanctum, I must step over bodies of gulls and rabbits that have been trapped here in the outer passage or else crawled in here to die. I do so carefully, averting my eyes as much as I can. After a while, spooked by the glare of the flashlight against stone, I switch it off, and let my eyes adjust. There's just enough light from where the heavy metal door hangs ajar to let me navigate the wide stone stairs and move deeper into the bowels of the old fort.

Once plastered white, the walls are marbled with grime, misted here and there in deep mold-green. Soon, though, it's too dark to tell. Despite stern words, spoken inward, I feel my pulse quicken.

At every corner, where the unknown looms tenebrous and forbidding, I need to force myself forward—take a breath, touch my fingers to the wall, feel my way around. I smell wet stone, soil, decay; the smell of the crypt. When it's hopeless to do anything else, I switch my flashlight back on.

It's true, then. I'm not alone. Or, not quite. Along the rough-hewn walls, my halo of light picks out first one dark body, and then another. I find three, clustered together, close to the ground, wings pressed tight like hands together in prayer. I have to get down in the dust on my hands and knees to look at them, to see the detail of them: the intricate adornment of the outer wings—a fretwork of ebony and sable through which faint threads of copper glint. Butterflies, still dormant. Soon to wake.

This is Inchkeith, an island in the Firth of Forth, just four miles across the water from Edinburgh. In its time, Inchkeith has been many things: the remote site for an early Christian "school of the prophets," later a quarantine island for those stricken with syphilis (banished there "till God provided for their health"), then a plague hospital and even an island prison, with water for walls.

So isolated was it, and yet so continually in view from the Scottish capital—as a rocky mirage upon the horizon—that it is said to have taken hold of the imagination of King James IV of Scotland, who saw in Inchkeith the potential for a notorious language-deprivation experiment. He, a polymath with a roving mind, was much taken up with concerns of Renaissance science, and practiced both bloodletting and the extraction of teeth. James sank huge sums into research into alchemy, human flight, and—according to a sixteenth-century chronicler—transporting to Inchkeith two newborn infants in the care of a deaf nursemaid, in the hope that the children, sequestered from the corrupting influence of society, would grow up to speak the prelapsarian "language of God."

Known as "the forbidden experiment" for the cruelty of inflicting such extreme isolation and irreversible social harm upon a child, its results were inconclusive. ("Some sayes they spak guid Hebrew," reported the chronicler, slyly, "but I knaw not." Others evoked a "brutish babble.") I suppose it depends what sort of God they were looking for.

In time, Inchkeith became a fortress isle, held sporadically by the English during times of war and, later—after great shedding of blood—the French. By the Second World War, this island half a mile in length was home to more than a thousand soldiers, with gun emplacements standing in tense watch over the entrance to the Forth. After Armistice, being too small and damaged and difficult to access to bother with in peacetime, the island was abandoned once more.

But as Inchkeith has fallen into obscurity, it has risen in environmental significance. Before the 1940s, only one seabird was known to nest there: the eider. In the decades since, it has become breeding ground to a dozen more, and hosted countless other visitors. By early summer, the cliffs will be heaving with life and limewashed with droppings, every ledge occupied with muddled nests of rotting seaweed or freckled eggs sitting directly on the stone, every species taking up its station in the strata of life: shags bedding down on the spray-spattered rocks; the sleek, monochrome guillemots on the lower reaches of the cliffs; the gnomish razor-bills, with their aquiline profiles on the next floor up; the elegant grayscale kittiwakes in the penthouse—and all of them shrieking in constant querulous protest at their neighbors.

Above, on what was once the lighthouse keeper's pasture, buxom puffins, with their candy-striped beaks, take up residence in warrens. Winter wrens and barn swallows and rock pigeons have taken ownership of the old military buildings which sag and split open like rotting fruit. Elder grows in thickets inside the roofless buildings, huddling together as if for warmth against the bitter wind that comes battering in off the North Sea.

As the weather draws in, gray seals heave themselves up concrete slipways grown slick with seaweed, to bask in weak sunlight—a thousand of them at a time, finding sanctuary enough here in the middle of a shipping channel to pup. Their spaniel-eyed babies spend the winter lolling in the tussocky grass, scooching up paths, and investigating the ruins. Around the same time, the butterflies and moths that blow over the island like smoke will begin to crawl into the dark tunnels that riddle the slopes to hibernate—blue-sequin peacock butterflies; the shining, shield-like herald moth; or scallop-edged small tortoiseshells like these. One twitches a leg. I leave them be.

I feel a draft, a faint movement that draws me upstairs. Far above, I see a glimmer of daylight. The faint alkaline taste of guano in the air. I find a door, half-rusted shut, but not past opening, and then I am out, standing alone at the very prow of the island, like the figurehead of a ship, looking out to see from the circular crater of what once was a gun turret, the last-ditch defenses of a war long over.

The wind is moving fast through empty space: mighty currents of the air, sucking the breath from my lungs. And the birds—the birds rise up as one great moving, wheeling mass. Screaming, squalling, outraged to see me—here, now, on this island of abandonment.

In this book we will travel to some of the eeriest and most desolate places on Earth. A no man's land between razor-wire fences where passenger jets rust on the runway after four decades' neglect. A clearing in the woods so poisoned with arsenic that no trees can grow there. An exclusion zone thrown up around the smoldering ruin of a nuclear reactor. A dwindling sea, upon whose deserted shoreline a beach has formed of the bones of the fish that once swam in its waters.

What links these disparate sites is their abandonment by humans: whether due to war or disaster, disease or economic decay, each location has been left to its own devices for years or decades. Over time, nature has been allowed to work unfettered—providing invaluable insight into the wisdom of environments in flux.

If this is a nature book, it is not one that rhapsodizes upon the allure of the untouched. This is, to an extent, from necessity. Fewer and fewer places in the world—if any—can lay claim to being truly "pristine." Recent studies have found microplastics and hazardous man-made chemicals even in Antarctic ice and deep-sea sediments. Aerial surveys of the Amazon basin reveal earthworks shrouded in forest that form the last remains of entire civilizations, long passed away. Man-made climate change threatens to transform every eco-system, every landscape on the planet, while long-lived artificial materials have etched our signature indelibly into the geological record.

There is no argument that some places are, relatively speaking, far less impacted than others. What draws my attention, however, is not the afterglow of pristine nature as it disappears over the horizon, but the narrow band of brightening sky that might indicate a fresh dawn of a new wild as, across the world, ever more land falls into abandonment.

Partly this is a reflection of changing demographics, as birth rates across the developed world fall and rural populations leave for the cities. Birth rates in nearly half of all countries have fallen below replacement levels; in Japan—where the population is fore-cast to fall from 127 million to 100 million or lower by 2049—one in every eight properties are already abandoned, forecast to rise to nearly a third of all housing stock by 2033. (The Japanese call them *akiya*, ghost homes.)

Partly too this is due to changing patterns of agriculture. Inten-sive farming—despite its many environmental drawbacks—is more efficient, using less land to produce more. Huge amounts of "mar-ginal" farmland, especially in Europe, Asia, and North America, is being allowed to revert to its wilder form. "Recovering secondary vegetation" (that is, former farmland and forestry) now accounts for around 7.1 billion acres, or more than twice the area of current cropland. It could rise to 12.8 billion acres by the end of this century.

We are in the midst of a huge, self-directed experiment in rewilding. Because abandonment *is* rewilding, in a very pure sense, as humans draw back and nature reclaims what once was hers. It has been taking place—is currently taking place—on a grand scale, while no one has been watching. This is, I think, an extremely exciting prospect. "The enormous and growing extent of recovering ecosystems worldwide," wrote the authors of one recent study, "provides an unprecedented opportunity for ecological restoration efforts to help to mitigate a sixth mass extinction."

While I have been writing this book, we found ourselves in the midst of a global pandemic. During this time, reports proliferated online of wildlife making forays into deserted streets around the world, while its human residents were confined in their homes. Marauding gangs of wild goats stormed the streets of Llandudno, Wales; sika deer grazed on traffic islands and stood on subway platforms in Nara, Japan; pumas stalked alleyways in Santiago, Chile; kangaroos bounded through the empty central business district of Adelaide, Australia.

Though they made for striking pictures, many of the most high-profile photographs featured animal populations already living on the peripheries of human settlement (the Nara deer, for example, are regularly hand-fed by tourists—and were likely roaming the streets in search of handouts). These were not so much examples of nature's *healing*, as nature finding the confidence to make itself seen. What they did remind us, however, was how tightly our own sphere of influence overlays, and intersects with, the nonhuman world, even now—and how quickly, therefore, spaces can be colonized by wild life should they truly be abandoned.

In the coming chapters I offer you the stories of twelve locations around the world, each of which embodies a different aspect of the process of abandonment and natural reclamation. These various locations, very different in climate, culture, and history, each offer their own flavor of melancholy and hope: they show us how every site, no matter how devastated, can come to recover in its own way, but also how human impacts can leave a long shadow for many years—decades—centuries—after these sites have fallen into disuse.

Some of my sites are literal islands; others simply act as them—enclaves of wildness in a sea of tarmac and brick or monocultural agricultural plains. The vast spoil heaps of West Lothian, Scotland, that we meet in chapter 1 were once described by the ecologist Barbra Harvie as "island refugia" for life, and it is this spirit that suffuses the rest of the book.

In part one, we will consider four sites emblematic of how the absence of humans allows wildlife to rebound—in some cases much more rapidly than one might expect. We will look at the basic processes of ecological succession, weigh the massive potential for carbon sequestration on abandoned lands, and consider how human crises such as war and nuclear disaster have produced exclusion zones that serve, effectively, as strict nature reserves—the absence of people, startlingly, proving more beneficial to an environment than contamination or minefields are deleterious.

Abandoned land, by definition, once belonged to someone. I expected humans to be present in the stories of these places only as a negative, but the more I traveled and researched, the more I realized that very few places are genuinely without their human occupants—whether they be holdouts from an earlier era, refusing to leave, or squatters who have moved in after the fact, seeking escape from the normal bounds of society or simply a place to stay. I came to see that this was a key plank of the story: the social and economic forces that drive abandonment, and the psychological forces acting upon those who remain—weathering the withdrawal of others, or else popping up in their absence.

To do without would be, as Henry James once said of his own ruin questing, a "heartless pastime." Inhabitants of sites of mass abandonment, most notably in the city of Detroit, have come to characterize the aestheticizing of their predicament—the presentation of its photogenic results without social context—as a form of voyeurism, or even "ruin porn." The human aspect, then, is what I focus upon in part two.

The neuroscientist David Eagleman once proposed that we have three deaths: the first at the point at which the body ceases to function, the second upon burial, and the third being "that moment, sometime in the future, when your name is spoken for the last time." Part three is an examination of a similar idea: the long shadow that we, as a species, leave upon the earth is an afterlife, of a sort. In this section, I visit places where our legacy persists even long after we are gone, places that make clear that it is not so simple as "we leave, and nature returns." We have written ourselves into the DNA of this planet, laced human history into the very earth. Every environment bears a palimpsest of its past. Every woodland is a memoir made of leaves and microbes that catalog its "ecological memory." We can learn, if we want, to read it—to observe in the world around us the story of how it came to be. In England, for example, one might spot the ghosts of ancient woods no longer extant by looking for shade-loving species like bluebells, wood sage, honeysuckle, creeping soft-grass—the flora of dappled glades now left stranded in gardens and on the shoulders of roads: indicator species pointing into the past. This memory, as with our own, affects how an ecosystem behaves in the present.

All of which leads us into part four, which offers studies of two abandoned sites that seem to me—and perhaps will to you too—to transcend their present and offer us a glimpse into a future in which climate change, and other human legacies, come to create a very different world.

I have spent two years traveling to places where the worst has already happened. These are landscapes wracked by war, nuclear meltdown, natural disaster, desertification, toxification, irradiation, and economic collapse. This *should* be a book of darkness, a litany of the worst places in the world. In fact, it is a story of redemption: how the most polluted spots on Earth—suffocated by oil spills, blasted by bombs, contaminated by nuclear fallout, or scraped clean of their natural resources—can be rehabilitated through ecological processes. How the hardiest ruderal plants can find their toeholds, colonizing concrete and rubble as they might sand dunes; how the palettes of ecological succession change as moss turns to golden grass, to the bright flags of poppies and lupins, to woody shrubs, to tree cover. How, when a place has been altered beyond recognition and all hope seems lost, it might still hold the potential for life of another kind.

PART ONE

IN ABSENTIA

The Five Sisters bing

❦ 1 ❦

THE WASTE LAND

The Five Sisters, West Lothian, Scotland

Fifteen miles southwest of Edinburgh, a knuckled red fist rises from a soft green landscape: five peaks of rose-gold gravel stand bound together by grass and moss, like a Martian mountain range or earthworks on the grandest of scales. They are spoil heaps.

Each peak rises along a sharp ridge from the same point on the ground, fanning outward in geometric simplicity. Along these ridges, tracks once bore carriages aloft, bearing tons of steaming, shattered rock: discards from the early days of the modern oil industry.

For around six decades from the 1860s, Scotland was the world's leading oil producer, thanks to an innovative new method of distillation which transformed oil shale into fuel. These strange peaks stand in monument to those years, when one hundred and twenty works belched and roared, wrestling six hundred thousand barrels of oil a year from the ground in what had been, shortly before, a sleepy, agricultural region. The process was costly and effortful, however. To extract the oil, the shale had to be shattered and super-heated. And it produced huge quantities of waste: for every ten barrels of oil, six tons of spent shale would be produced. In all, two hundred million tons of the stuff—and it had to go some-where. Hence these enormous slag heaps. Twenty-seven of them in all, of which nineteen survive.

But to call them slag heaps is to understate their size, their stat-ure, their constant presence in the landscape; unnatural both in form and scale. Locally, they are called "bings"—from the Old Norse, *bingr*, a heap, a tip, a bin.

This particular formation, the five-pronged pyramid, is known as the Five Sisters. Each of the sisters slopes gradually to its high-est point, then falls steeply away. They rise from a flat and other-wise rather unremarkable landscape—muddy fields, pylons, hay bales, cattle—to become the most significant landmarks of the re-gion: some pyramidal or square; some organic and lumpen; others still rising raw-flanked and red to plateaus like Australia's Uluru.

Mere tips at first, they grew into heaps that shifted and re-formed like dunes. Then hillocks. Then, finally, mountains made from small chips of stone—each the size of a fingernail or a coin, with the brittle texture of broken terra-cotta. These mountains grew and spread, as barrow after barrow was dumped upon the heap. They rose from the land like loaves, swallowing all they came into contact with: thatched cottages, farmyards, trees. Under the north-ernmost arm of the Five Sisters an entire Victorian country house—

stone-built and grand, with wide bay windows and a central cupola—lies entombed beneath the shale.

Oil production continued on a massive scale here until the Middle East's vast reserves of liquid oil came into ascendancy. In Scotland, the last shale mine closed in 1962, bringing to an end a local culture and way of life, leaving mining villages without the mines to employ them, and only the massive, brick-red bings as souvenirs. For a long time the bings were disliked: barren wastes that dominated the skyline, fit only to remind the region's inhabitants of an industry gone bust and an environment pillaged. No one wants to be defined by their spoil heaps. But what to do about them? That wasn't clear.

A few were leveled. A few later quarried afresh, as the red stone flakes—"blaes" as they are technically known—found a second life as a construction material. For a time blaes turned up everywhere: fashioned into pinkish building blocks, used as motorway infill, and—for a time—surfacing every all-weather pitch in Scotland, including the one at my high school. Blaes stuck in grazed knees, collected in our gym shoes, left a telltale dust across the sweaters used as goalposts—and generally formed the brick-red backdrop to our communal coming of age. But mainly the bings lay abandoned and ignored. After a while, the villages in their shadows grew used to their silent presence. To enjoy them, even.

It's easy to find the bings. You can see them from miles off. Just drive until you can't get any closer, and hop the fence. There's no fanfare. They are spoil heaps the size of cathedrals or hangars or office blocks, rising from the fields in artificial formations.

My aunt and uncle live in West Lothian, not far from the Five Sisters and even closer to their even larger cousin at Greendykes. Last time we went to visit my relatives, my partner and I took a detour to climb the sleeping giant. The light was flat and silver, the sky gray and cottoned over with cloud. We parked in a semi-derelict industrial estate, between rust-streaked Nissen huts and faded signposts, and wandered out into a landscape of almost unbelievable strangeness, like the first colonists on a new planet. Sculpted by wind and rain, there were outcrops and boulders comprised of a conglomerate of compressed blaes, a rock form all its own, in Martian red and violet-gray where the outer blaes had chipped away to reveal fresher stones—with that smooth, almost greasy look of chipped flint, olive-tinged—not yet discolored by oxidation.

Deep ponds of bottle green had gathered in hollows at the base of the slope, at the foot of each dell and gully formed by the tip's puckered edges, their outlines picked out in the acid green of the pond weed and hair-thin grasses that intermingled in the shallows. Water lilies poked their noses through the surface, where tiny insects skated by. Whip-thin birches sprung with unlikely fervor from their gravel beds, silk-skinned and shining and bearing tiny buds of sweet new leaves. We pressed between the birches, along a narrow footpath, to emerge at the base of the bing proper, and found its vast red flanks rising ahead of us, contours and crannies picked out dramatically in vegetation, and striated with tracks.

We began to climb, but the going was difficult. The blaes had solidified into a dense conglomerate to form rock faces in places, in others scree. Elsewhere, the outermost layer was grassed over but crumpled, like laundry, where the skin had slipped down, and when we put our weight on it we post-holed as if through rotten snow. Grit collected in our shoes. We had to stop to empty them out, and I felt a flush of something like nostalgia.

After a fashion, we reached the top—a wind-battered upland that offered panoramic views across clean-swept fields to Niddry Castle, a sixteenth-century tower, behind which yet another bing—a sheer cliff of spent blaes, ruddy-faced but streaked with green and gray—stood breathing down its neck. And beyond, yet more, rising proud from the flats.

The flora here was a strange mix; it was hard to get a fix on the sort of climate we found ourselves in. Russet shoots of fireweed were coming up across the tops, as it might along any roadside in the country. But other than that, the vegetation had a sparse, sub-arctic feel: a close crop of soft-furred leaves and starred flowers and short, blond grass. But there was red clover too, with their sweet heads full of nectar just beginning to open, and spotted orchids. The year's first bumblebees blundered by, revving their engines. Buds and shoots were snaking up, out of the gravel. The land was basking, warming, ready to bloom. It was the end of April. It was impossible not to think of T. S. Eliot:

breeding
Lilacs out of the dead land, mixing

Memory and desire, stirring
Dull roots with spring rain.

Back in 2004, the ecologist Barbra Harvie made a survey of the bings' flora and fauna and found to almost everyone's surprise that, while no one had been looking, they had transformed into unlikely hot spots for wildlife. "Island refugia" she termed them: little islands of wildness in a landscape dominated by agriculture and urban development. Hares and badgers, red grouse, skylarks, ringlet butterflies and elephant hawkmoths, ten-spotted ladybirds. Among the flora were a diverse array of orchids—the vanishingly rare Young's helleborine, a delicate, many-headed flower in pale greens and pinks, found in only ten locations in Britain (all post-industrial, two of them bings); the early purple orchid in ragged mauve; the greater butterfly orchid, with its winged petals—and a genetically distinct birch woodland that had established naturally at the base of the tiny bing at Mid Breich.

Overall, Harvie recorded more than three hundred and fifty plant species on the bings—more than can be found on Ben Nevis, the UK's highest mountain—including eight nationally rare species of moss and lichen, among them the exquisite brown shield-moss, whose thin tendrils loft targes to the sky like an army in miniature. Over the space of a half-century, these once-bare wastelands had somehow, magically, shivered into life.

Eliot's wastelanders—or some of them—transpire to be his contemporaries: modern commuters flooding across London Bridge at dawn, lonely typists whiling away evenings in studio apartments. In a sense, we are all residents of the Waste Land still—and I felt it keenly then, standing at the prow of this great memorial to ecological degradation.

> What are the roots that clutch, what branches grow
> Out of this stony rubbish?

Eliot's *Waste Land* drew from the "perilous forest" of Celtic mythology, a land "barren beyond description" through which a hero must pass to find the Otherworld, or the holy grail. The bings too already offer a glimpse of what we might find on the other side: recuperation, reclamation. A self-willed ecosystem is in the process of building new life, of pulling itself bodily from the wreckage. In starting again from scratch, and creating something beautiful.

Blasted to 950°F before they were dumped, still roasting, on the tip, the blaes would initially have formed a vast sterile desert devoid of seeds or spores. The regrowth we see now, then, began from absolute zero—no soil, no nothing—as part of a process known as "primary succession."

First came the pioneers: lacy foliose lichens, curling at the edges and growing in coral-like reefs; *Stereocaulon*, the snow lichens, forming up in crusts. Green mosses laid over the gravel like a picnic blanket, soft and welcoming. Then, the ruderal plants—from the Latin, *rudera*: of the rubble—the wildflowers and deep-rooted grasses that colonized the loose chutes of scree, stabilizing them like marram grass on sand dunes. Kidney vetch and toadflax, bluebells and plantain, yellow rattle, pearlwort, speedwell, sweet cicely. In the damp clefts, seeds of the hawthorn and the rosehip and the birch caught purchase, took root.

All these materialized as if by magic: blown in on the winds, or spread by birds, or left in the droppings of animals (what ecologists call, poetically, "seed rain"). They are the few survivors of a much greater experimental program, the hardy few who found a toehold in the spoil heaps and made it work for them. The more there are, the easier it becomes for others, as organic matter builds up as leaf mold and deadwood and algae, and acts as a compost for the next generation. To begin with, the bings would have been species poor, and then a fluctuating assemblage of species would have played across their faces as each tried out new forms of what they might become. Montane species, common weeds, escaped ornamentals. But over time, species accrue, bed down. And now, the bings come to act almost as an archive of biodiversity for the local area.

And though the bings are a remarkable example of primary succession in action, they are not unprecedented. In nature, the process takes places only rarely—on newly formed dunes, or volcanic islands bubbling into the open from underwater vents. But humans have a bad habit of stripping the land bare of all its life, and starting the process all over again.

In the wake of the London Blitz, the then-director of Kew Gardens noted a similar process taking place in the charred and ruinous bombsites that pitted the capital. In a 1943 pamphlet, "The Flora of Bombed Areas," E. J. Salisbury described "the rapid clothing of the blackened scars of war by the green mantle of vegetation." These plants grew up spontaneously, he noted, upon the bare rubble and in the ruins of the houses. The "dust-like spores" of mosses, ferns, and fungi drifted in through the broken windows; the soft, silken seeds of the fireweed parachuted in from site to site (each young plant, he added, might produce eighty thousand seeds a season). So too did the pennant-yellow flags of ragwort and groundsel and coltsfoot, and the wispy, wand-like fleabane, and the sow-thistle and the dandelion, and the tiny, star-flowered chickweed.

All the time, these seeds and spores—the potential for wildflowers, of wild life—is drifting by us on the air, waiting for their chance. As a petri dish left out will soon grow cultures of its own, a sterilized bombsite or lava flow or bing will do the same, but on a grander scale. All the seeds need is a place to land.

And as the wounds of the Blitz were skinning over in London, and the oil shale industry was sputtering to a close in lowland Scotland, on the other side of the world a similar process was just cranking into motion in the wake of yet more bombs. This time, underwater.

The Bikini Atoll, a ring of coral islets encircling a turquoise lagoon, was used by the United States as a nuclear weapons testing site during the 1940s and '50s—most notably for the 1954 Castle Bravo test, when a thermonuclear device more than seven thousand times the force of the bomb dropped on Hiroshima, producing an explosion of such unexpected force it shocked the scientists that designed it and ultimately prompted a worldwide ban on atmospheric testing.

The blast gouged a crater more than a mile across and 260 feet deep, vaporized two islands, and formed a vast mushroom cloud of steam, superheated air, and pulverized coral, a luminous globe of fire like a second sun, and turned the sky scarlet. It rose 130,000 feet into the atmosphere, before the fallout snowed back down upon the Marshall Islands in a blizzard, burning everything it touched. The waters of the lagoon flash-boiled as temperatures rose to 99,000°F, and rushed outward as waves 100 feet high, which stirred up a million tons of sand that smothered any coral that had survived the initial explosion. It left a barren underwater wasteland, grossly contaminated and utterly devoid of life.

But in 2008, when an international team of researchers returned to the atoll to inspect the lagoon, they found to their surprise that a thriving underwater ecosystem had formed in the blast crater over the intervening decades. It looked, as one coral scientist commented in wonder, "absolutely pristine." While above water the island remained eerily abandoned—uninhabited except for the caretakers of a tiny tourist initiative*—and its groundwater and coconuts unfit for human consumption, the lagoon below was a whirl of kaleidoscopic life. Less so than before—twenty-eight species of coral were still missing—but, nevertheless, now as one of the most impressive reefs on the planet, where corals grew as huge rocky cushions the size of cars, or as dendroids twenty-six feet tall, with slender branching fingers.

A team from Stanford University dived the crater again in 2017 and found it was even more densely embroidered with life. Hundreds of schools of fish—tuna, reef sharks, snappers—flashed through limpid waters. It was, reflected project lead Professor Stephen Palumbo, "visually and emotionally stunning." In a strange way, he said, the new reef had been protected by the atoll's traumatic history—as a direct result of the lack of human disturbance, the fish populations were bigger, the sharks more abundant, and the coral more impressive.

From the embers a cornucopia of life has risen. Here it was not carried on the wind or by birds, but ocean currents. Coral larvae—the dust motes of the sea—are thought to have swept here from Rongelap Atoll, seventy-five miles away, and begun a new colony on what was then a cratered moonscape dusted with the talc-like remnants of their predecessors.

---

*The 167 native Bikinians, asked to leave their home "temporarily" for the US tests ("for the good of mankind and to end all world wars"), were allowed to return in the early 1970s only to be evacuated once more in 1978. Life on Kili, another atoll in the Marshall Islands where many exiled Bikinians now live, has been very challenging; it lacks a natural harbor, and is increasingly impacted by extreme tides and flooding. In 2015, the Marshallese foreign minister described it as now "uninhabitable" due to climate change.

Again—this latency of life. It drifts around us all the time, invisible, like an ether. It's in the air we breathe, the water we drink. Savor it: each breath, each sip, is thick with potential. In this cup of nothing is the germ of everything.

The self-seeded ecosystems that have emerged on the bings—and on derelict sites like them—tell us a great deal about the possibilities and process of natural recovery; about nature's resilience and capacity to recover after what would seem like a death blow.

These are stories of redemption, not restoration. These sites never again will return to the way they were. But what they do offer us is insight into the processes of reparation and adaptation, and—more valuable still—they offer us hope. They remind us that, even in the most desperate of circumstances, all is not yet lost.

And there is a great deal we can learn from them. Indeed, there has been a sea change in how post-industrial, or other "anthropogenic," sites are perceived and valued in recent years. Some of the most exciting developments in ecology and conservation have been in the study of landscapes deeply impacted by human activity; in observing how ecosystems might expand and contract, adapt to new conditions, take a heavy blow but come up fighting on the other side.

Some of the new foci of scientific interest are sites that, on first glance, might be dismissed as drab or run down or ruinous: to appreciate their significance requires a certain retuning of the eyes and adjusting of sensibilities when we look at the world around us. It is much harder to recognize the value of lead when it sits so pale against the flash of silver or gold. But these *terrains vagues*, with their self-willed communities of hardy plants, may be more authentically *alive*, more solidly *real*, than many of the world's most celebrated beauty spots—and in that way offer an attraction and a value all their own.

Some of the earliest work on how to evaluate the scrappy, self-made ecosystems that spring up in abandoned places took place in post-war Berlin where, as in London, large tracts of urban land had been left as rubble and ruin by air raids. But, unlike in London, reconstruction here was retarded by the construction of the Berlin Wall and the division of the city. Rail yards in West Berlin, for example, fell silent after East Germany rerouted trains to avoid the Allied-occupied zones.

In the Tempelhof shunting yard, with the site in stasis, nature began the reclamation. The tracks remained, but broad-boled birch muscled up between the sleepers, blocking the tracks and halting the return of the trains. A complex mosaic of grassland, shrubland, and groves of black locust trees sprung up under a rusting water tower; by 1980, the 25-acre Natur-Park Südgelände supported 334 species of ferns and flowering plants, plus foxes, falcons, three hitherto unknown species of beetle, and a rare spider previously only found in underground caverns in the south of France.

Ingo Kowarik, a local ecologist, made a detailed study of the site and—based on his findings there and in similar abandoned sites across the city—devised a new framework through which we might begin to understand their significance. In all, he wrote, there are four different types of vegetation. First, the remnants of what we might considered "pristine" nature—ancient woodland and other undisturbed sites. These sites are very valuable, being highly diverse and densely structured. Next, cultural landscapes—that is, where nature has been shaped and sculpted by farmers and forest-ers. Third, the trees and plants that have been added for ornamental reasons, an esthetic element of urban planning. Then, finally, what Kowarik memorably classifies as "nature of the fourth kind": the spontaneous ecosystems that have grown up on wasteland, unsup-ported. His point is that these new feral ecosystems, in their au-thenticity and self-direction, are a new form of wilderness worth preserving in their own right.

In Britain, a similar story unfolded in Canvey Wick, where a 230-acre patch of land was used first as a dump for sediment dredged from the shipping channels of the Thames, and later de-veloped as an oil refinery. Huge circular pads of concrete were laid in preparation for the installation of outsized metal holding tanks, but building stalled during a crash in oil prices and the site never came to fruition. It was considered an eyesore until, in 2003, entomologists identified dozens of rare invertebrates living there, including three hundred species of moth, and insects so rare they did not have English names. Surveys later revealed the site to have more biodiversity per square foot than any other site in the UK. It was, one conservation officer extoled, "a little brownfield rainforest"—and became protected as a Site of Special Scientific Interest in 2005.

A few months ago, I visited a comparable brownfield wonderland, closer to home: the Ardeer peninsula on Scotland's southwest coast, once a vast complex of sand dunes and salt marsh that became a cradle of industry in the nineteenth century when Alfred Nobel constructed a dynamite works and testing ground along its remote stretch of coastline. At its height, the site employed thirteen thousand people in its laboratories and production lines, and stored nitroglycerine in thousand-gallon tanks. Buildings were built well apart, nestled behind embankments sculpted from the sand hills in case of accidents. (For there were accidents: in 1884, ten local girls filling dynamite cartridges were killed in a massive blast; "not a vestige of the hut remains," reported the local paper. Parts of the body of one of the girls was found more than 150 yards from the scene of the explosion.)

Those huts now lie tumbledown and open to the elements, the blast walls grown over with heather. Old paint flakes onto the floor and gather in drifts with the fallen leaves. Faded signs warn: DANGER—EXPLOSIVE ATMOSPHERE.

Iain Hamlin, a local conservationist campaigning against the site's redevelopment, led me through a gap in the fence onto the rail platform, which stood eerily amid a clearing in the trees as if awaiting the last train. The old parking lot was an open expanse quilted with soft brown moss and frothy gray and peppermint lichens, which seemed to shimmer like the surface of an impressionistic pool, whipped up in some places and lying still in others.

Tussocks broke through its smooth surface, and goat willow hung heavy with tasseled catkins. Sea buckthorn had pushed up along the seams, its burnt-orange fruits now sagging unpleasantly on the branch, bleached to a sickly pallor—food for the birds. When I scratched my heel into the spongy matter underfoot, it parted to reveal the crumbling asphalt underneath, like bone. Iain dropped to his knees to point out the pinhead tunnels of the minotaur beetle, which rolls rabbit droppings down into underground pantries, and the telltale burrows of solitary bees. Farther back, cooling ponds strewn with rusted pipes were busy with teals and moorhens. An old concrete streetlight stood incongruously in the woods beyond: some ravaged Narnia. Jays catcalled overhead.

Though profoundly altered by development, the sites at Ardeer and Canvey Wick are almost uniquely well suited to become hubs of biodiversity. Old concrete and tarmac hinder succession—keeping the ground clear of forest, which can, somewhat counterintuitively, inhibit biodiversity rather than improve it—and opens the area to the light. So too do the roving local teenagers, whom we saw setting brush fires and clambering onto the roof of the derelict power plant. The combination of so many miniature sub-habitats in close proximity is an ideal situation for many insects, which have different requirements at different stages of their life cycles. Derelict buildings—strangely beautiful in their slow decay—also offer hiding places for hibernating butterflies and moths, whose chrysalides and cocoons have been spotted in their hundreds there hanging on dank, dark walls.

Given the intensity of contemporary agriculture—where monocultural swaths can stretch away to the horizon—it is becoming increasingly well recognized that ruinous, utterly neglected sites such as these have become refugia for wildlife; indeed, according to the conservation trust Buglife, "the invertebrate rarity and diversity of some brownfield sites is only equaled by that of some ancient woodlands." A remarkable feat, given that most brownfield sites have usually been in existence for only a few decades—when a woodland might take hundreds of years to come to full maturity and ecological complexity.

As a result of these findings, there has been a sea change in the way we look at the ecological world around us. Consider this: back in the seventeenth century, the term "wasteland" was often applied not to sites of dereliction but to fens, swamps, and marshes. These regions were seen, essentially, as wastes of space—unkempt ground, ill-suited to agriculture, difficult going for travelers—and were targeted for "improvement," so as to turn them into productive farmland. Now, seventeenth-century "wastelands" are considered invaluable wetland ecosystems bustling with rare species, which also play significant roles in flood control and carbon sequestration. Millions of dollars now go into their preservation, and into the blocking up of old drainage ditches.

The lesson in this, I think, is that what feel like self-evident truths about the world around us can, in fact, be culturally specific—moral judgments we are imposing upon the world around us. If we want to do the best for the environment, what we need is a new way of seeing: a new way of looking at the land.

In densely populated, intensively managed regions like the UK and Europe, some of the only places growing truly wild and un-regulated may be those which have already been used and then discarded. Contrast the disheveled wastes of Canvey Wick—where bugs curl inside untrimmed stalks for winter, rare spiders lurk in damp heaps of fallen wood, and adders bask on pavements warmed by the sun—with a garden, primped and preened, high-maintenance yet skin-deep.

What eyesore sites like the wastelands can teach us is a new, more sophisticated, way of looking at the natural environment: not in terms of the picturesque, or even the care with which it has been tended, but with an eye upon its ecological virility. After learning to do so, the world looks very different. Sites "ugly" or "worthless" on first glance can transpire to be deeply ecologically significant—and their ugliness or worthlessness might very well be the quality that has kept them abandoned, saved them from redevelopment or overenthusiastic "management"—and, therefore, destruction.

Aldo Leopold said that our ability to perceive the quality in nature begins "as in art, with the pretty." After that, it expands "through successive stages of the beautiful to values as yet uncap-tured by language." What he was getting at is this: knowledge deepens appreciation. Leopold saw a marsh, veiled in a thin gauze of mist, aglitter at the low light of dawn, watched cranes descend upon their feeding grounds in "clangorous descending spirals"—and saw not only this, but the history of the cranes and all their evolutionary predecessors who had spiraled down upon this wet-land, and all those like it, over eons; he held to his eyes like binoc-ulars an understanding of how this momentary pastoral scene formed a necessary component to, or a synecdoche of, a greater, wondrous whole.

This too is a form of beauty—a conceptual one, in the way that mathematicians might come to appreciate a particularly elegant equation, or an artist might consider an empty room lit only by a flickering light, or half-filled with crude oil, and feel staggered by its unnerving implications.

And, as with other forms of aestheticism, it can be taught. To come into an abandoned mine or spoil heap or quarry or parking lot or oil terminal, and see it for the natural wonderland it has become is, I admit, a difficult ask. But in these environmentally straitened days, it is a taste worth cultivating.

As wetlands were once drained in the name of progress, similar miscalculations took place in West Lothian. In a follow-up to her floral survey, Barbra Harvie studied the effects of "management" methods employed on some of the remnant bings; from the 1970s, efforts were made to ameliorate the appearance of some of the heaps using invasive "restoration" methods: peaks and ridges were rounded off, topsoil was imported, a commercial ryegrass mix was sown across their faces. This was done, she noted, for entirely esthetic reasons—to make them look more "natural."

But these efforts failed. Nutrients from the new topsoil leached away within a few short years, and the planted species died. With-out constant treatment with fertilizers, the managed bings grew

bare and barren—far worse off than those left to their own devices. Species poor, nutrient poor—these "managed" sites offer a
cautionary tale about prettification.

A similar argument might be made about the High Line of New
York—a former raised railway, which grew thick with spontaneous
vegetation after it fell into disuse. Now it has been transformed
into a popular public space, a park only thirty feet wide but a mile
and a half in length. When I went there in person, however, I
found—to my surprise—that the old, self-regulating greenery had
been dug up and replaced with a maintained garden "inspired by"
the original, rambunctious self-seeded community. I found a clar-
ification on the website later: "Naturalistic plantings try to recre-
ate the emotional experience of being in nature. While the gardens
may feel natural, they are anything but. Plants that would never
meet in the wild are planted together here in manufactured soil.
The gardeners weed, water, prune, edit, and shape. Human intervention is a constant and many conditions of the site are artificial."

Sites like the High Line can be valued in different ways—but,
environmentally speaking, this curatorial impulse of ours—so intrinsic to the way we think about the world, so deeply ingrained in
Western culture—is a damaging one. (In 1967, the historian Lynn
White Jr. argued that the roots of our current ecological crisis can
be traced back to the Judeo-Christian "arrogance" toward nature.
In Genesis, God awards man dominion over all nature—its birds,
its fish, its cattle, "every creeping thing that creeps over the Earth."
"Especially in its Western form," White added, "Christianity is the
most anthropocentric religion the world has ever seen.")

Though we enthusiastically embrace our self-appointed role as steward of the planet—pruning here, planting there, tidying messes, and getting "pests" under control—we are not always successful. Gardens, parks, and farmland are often ecologically dull, their continued existence precarious and dependent upon our benevolence, while hedgerows, road shoulders, and the *terrain vague* of cities might be vibrantly biodiverse and deeply rooted. We weed out plants well suited to the ground and conditions, and insist on propping up expensive, ill-suited, ornamental ones. Better, perhaps, to resist the impulse. Step back.

Marcel Duchamp was talking about art when he declared that "esthetic delectation is the danger to be avoided." But there is something universal in it too: it's not that these sites are *not* beautiful, but that our eyes are not yet trained to appreciate them for what they are and what they symbolize. Instead our eyes are snagged, seduced by the flimsy suggestion of plenty.

I think of it as a distinction not unlike that made between the doe-eyed aspirational "commercial" catalog models and the angular, even awkward, look of high-fashion subjects; photographers are often drawn to faces that might be classified under that difficult-to-translate French class of the *jolie laide* (literally, "pretty-ugly," a term applied to women whose imperfections elevate them from conventional attractiveness to a higher plane of visual interest). The bings, and places like them, might be a jolie-laide landscape: ones whose industrial scars only serve to throw their current stature and ecological significance into sharp relief.

In 1975, the avant-garde conceptual artist John Latham was

tasked by the Scottish Development Agency with reimagining these giant spoil heaps—then considered a blot on the landscape—and finding them new purpose. Rather than recommend their reshaping or removal, he praised their "immaculate and classical nature," and insisted instead they simply be retained and reconceptualized as "process sculptures."

In support of his proposal, Latham produced satellite images of the constellation of bings of which Greendykes forms a part, declaring them elements of a giant "Niddrie Woman"—a massive work of land art constructed by ten thousand hands over decades, along the lines of ancient hill figures like the Cerne Abbas Giant or the Uffington White Horse, in a "modern variant of Celtic Legend." While hiking up Greendyke bing, then, we were traversing her great bare belly; her head, as Latham envisaged, rose up as the Albyn Works bing to our south, beyond the chasm of her clavicle, in which a stagnant green loch was crowded with waterfowl. Hopetoun bing, flung to our north, served as her disembodied arm, while those red cliffs at Niddry comprised her oversized heart.

It was a clever sleight of hand, this conceptual rebrand. To the government, not much was required beyond, perhaps, their preservation. Niddrie Woman was a *cheap* solution to the bing question, if one happened to be looking at it that way. So, soon enough, the Greendykes bing was scheduled as a national monument—one reason it has been spared the bulldozers long enough to be recolonized by nature.*

---

*When Latham died in 2006, his ashes were scattered in West Lothian, in the heart of the Niddrie Woman.

As we staggered across the tops, harried by the wind, I tried to look at her with less critical eyes. I saw afresh the sculptural form of her torso, the dark clean lines where dirt bikes had worn parabolic paths in the curve between body and head—passing again and again along the purest lines of her form like a charcoal drawing. I looked at the mottled colors of the blaes themselves, which ranged in tone from coral to cantaloupe, flecked through with the soot-blue of the original, unfired oil shale. I looked to the little flowers that frilled the path's edge, the rosaline lichens. I opened my ears to the murmurs of insects and the piccolo trill of a skylark in ascendance. Niddrie Woman is a process sculpture in more ways than one. She is the "unconscious sculpture" of the oil shale industry, as Latham intended, and as they are officially recognized. But she is also a monument to the process of succession, recovery, reclamation.

In his "feasibility study," Latham compared Niddrie Woman to the *Venus of Willendorf*, a Paleolithic statuette with pendulous breasts and capacious belly, thought by some to have served

as a fertility symbol. Yet surely there can be no better symbol of fertility—of the victory of fresh life over sterility—than the bings themselves.

What rite of spring might one conduct in a place like this? Days later, on the last night of April, I attended the fire festival Beltane, on Edinburgh's Calton Hill, a ritualistic performance of the death and rebirth of the "Green Man," a mythical Celtic figure representing the cycle of fresh growth each spring, and his courtship of the mother goddess. Around them, cavorting figures—painted scarlet and buck naked but for a loincloth—drummed and writhed and breathed fire and mewled miserable wails. A wild bacchanal, in which participants might lose themselves, shrug off old inhibitions. That, perhaps. Or think instead of the Grauballe Man, the bog body found in Jutland, his throat slit and his stomach stuffed full of seeds: clover, ryegrass, fat hen, buttercup. Latham, I'm sure, would approve of either.

On our way back down, sliding through the blaes on our heels and hands, we crossed paths with a dirt-biker, who roared up and down the slopes of scree like Evel Knievel, defying gravity. He wore a helmet with a mirrored visor that he never removed, and though he paused between runs to allow us to scamper out of his path, he didn't speak or make any other signal to indicate he was aware of our presence.

It was an unsettling encounter. Several times I had to step off the compacted path and let my feet sink into the loose blaes, wading as if through deep snow. As we reached the edge of the green loch, a red doe burst suddenly from the undergrowth and made a break for it up the slope toward the biker. She was fully grown, sleek and russet, with powerful back legs that pedaled for purchase against the loose stones. For a long time she seemed to be losing height almost as fast as she gained, but we stood stunned, all three of us, watching for several minutes as she desperately scrabbled, sending rose-gold shale clattering back down the slope, until at last, finally, she passed the mute biker, rounded the first false peak of shale, and disappeared into the wild land beyond.

✿ 2 ✿

NO MAN'S LAND

The Buffer Zone, Cyprus

Yiannakis Rousos can see his house from here. I bring my head down so my eyes can follow where his finger points across the fields to a square, two-story block that stands alone down by the sea. It was his family home. It and all the land between where I stand and the coast, around twelve acres of it, is his still—or at least he has the title deeds.

Back then, all this was citrus orchards: oranges, grapefruits, and lemons hanging heavy from the branch, piled high in the basket. His family was rich, he says. They had everything. Then, one day, nothing.

In the early hours of July 20, 1974, Turkey invaded Cyprus in the aftermath of a Cypriot military coup. The coup had been the culmination of decades of tension and sporadic violence on the island between the ethnic Greek and ethnic Turkish communities, and—backed by the junta in Athens—intended to force enosis, or union, between the island and Greece.

Though the shock arrival of the Turkish troops was billed as a "peace operation," more than three thousand people were killed in the fierce fighting that followed and several thousand more are still missing, presumed dead. More than a third of the island was, and remains, occupied.

While the Turkish forces were taking control of swaths of the island republic, one hundred and fifty thousand Greek Cypriots fled their homes, the Rousos family among them. When they left, they left at a sprint. They were running for their lives. What they didn't realize was that they were leaving their lives behind them.

Two days after their departure, their farm still overrun by Turkish forces, Yiannakis's father returned alone under the cover of darkness, slinking through the orange groves that he knew so well, vanishing between the trees as soldiers passed by on their patrols. He slipped into the house unseen and, hands shaking, searched the family papers for their most important documents. In the yard outside, all the animals except for a dog and a single pig were dead.

Four weeks later, a ceasefire was declared, and the Turkish troops stopped from making further advances. The two warring factions were physically separated by means of a tightly policed demilitarized zone that stretched the island end to end, containing the Turkish troops in the north of the island and the Greek Cypriots in the south. This corridor was 112 miles long, and anywhere from 11 feet to 4.6 miles wide. It blocked off roads, encompassed entire villages, and sliced the capital, Nicosia, in two. It also cut through the Rousos's land, leaving their house on the other side.

Over the following months, the family watched from a distance as the citrus trees, which they had so carefully tended and watered, dried out and died. Today the house is derelict and the land repurposed as wheat fields by unknown Turkish farmers. Sometimes, says Yiannakis, he can't sleep for thinking about it. We stand together at the very edge of no man's land, the buffer zone between the Republic of Cyprus and Turkish Republic of Northern Cyprus—a state whose official existence is recognized only by Turkey, and certainly not by Yiannakis Rousos—the way forward blocked by three decks of razor wire, stacked on top of one another and rusted solid. The "wire of shame" as Yiannakis calls it, which stretches from here on the east coast to the western shore.

Not far from where we stand rises a two-story observation platform flying both the Greek and the Cypriot flags. Perhaps a hundred meters away, its equal and opposite, with the Turkish flag and its negative, the red-on-white crescent moon of North Cyprus. The buildings seem to square off, and inside can be seen the shadowed figures of watchmen holding binoculars to their eyes. The soldiers change every shift, but the routine stays the same, as it has every day, every month, every year, since 1974.

In the meantime, the Rousos family grew up, grew old. Their changed circumstances—one day wealthy, the next, penniless—was hard on them all, but especially on Yiannakis's mother. She couldn't adjust to life as a refugee, with "no food, no money, no clothes, no shoes." She suffered terribly from intense emotional strain. She spent time in the hospital. It's partly why, Yiannakis says, he never had a family of his own. He wanted to wait until they had their land back. These days, with property on the island at a premium, a coastal property like theirs would make him a very rich man once more. Until then, he adds, "I can't even sleep in my own house."

The spring air is warm, a sea breeze bundling over the fields. Clouds hang threateningly overhead, the sky bright and dark all at once. Yiannakis picks an orange blossom from the garden of the last Greek-Cypriot house before the fence line; it has a thick, heady scent. The flowers have just opened today, he says. You notice that sort of thing when you've grown up on a citrus farm. On the lemon trees, fruit already hangs heavy and green. I can smell them from here: fresh and clean as disinfectant.

We turn away and walk along a road that abuts the edge of the buffer zone. Stray cats wander the track ahead of us, their fur matted and wet. The nearest fence bears small triangular red signs, warning of land mines in three languages. He thought they'd be back later that afternoon, Yiannakis says. Soon, he had to revise his expectations: a few weeks' time. Then, a few months'. It's been forty-five years. "Maybe," he says, with a look of resignation that borders on amusement, "in twenty years I will be telling the same story."

What began as a stopgap in wartime—a desperate peacekeeping measure—an abstraction drawn in green pencil on a map, has become etched into the very fabric of the land. Rebecca Solnit once wrote of the "blue of distance," the color of hills that recede layer upon layer unto the horizon. Well, this is the green of time. The green that grows from nothing, anything, if left for long enough. It comes at first as mildew or mold. A misting of green-gray, or mustard-green, the green of decay. But then it grows and grows into the verdant palette of new life: leaf green, lime green, the green of fresh new shoots.

Over time, where no man could step without risk of arrest, or painful death, or international crisis, other life slowly took hold. Nasturtiums curled through the cracks in the pavement. Cacti tumbled from balconies. Palms sprang up in the middle of the roads. Trees and bushes began thin and far apart, then grew thick and strong and joined forces. Each one became a timer set ticking, a marker of time passed in bloody stalemate.

Now the stalemate has taken solid form, clearly visible on satellite imagery: the broad-stroked sweep through the remote west, the delicate emerald stitching through Nicosia Old Town, the curling scrawl east out to the coast. The green line made real.

Years ago, on the other side of the world, a group of explorers noticed something strange. It was the turn of the nineteenth century, and two American army officers—Captain Meriwether Lewis and Lieutenant William Clark—had been tasked with surveying and mapping the territory recently acquired by the United States under the Louisiana Purchase. Between May 1804 and September

1806, Lewis and Clark's Corps of Discovery expedition crisscrossed the interior of the continent between St. Louis and what is now Astoria, Oregon, in search of a watercourse through the new territory to the Pacific.

During this period, the corps passed through Native American territories and remote backcountry regions. Their men had to get by on the spoils of hunting and were thus keenly aware of the availability of game, recording every buck and rabbit they killed. Throughout the spring of 1805, they had enjoyed a particularly fruitful period as they tracked the Missouri River upstream. In what is now Montana, they found an untouched wilderness, practically overflowing with wildlife.

It was an Arcadian scene. Geese landed in vast flocks upon the water meadows; buffalo, elk, and antelope grazed in herds so huge they stretched off in all directions. It was "beatifull in the extreme," Lewis observed in his journal. The animals, he added, were "extreemly gentle the bull buffaloe particularly will scarcely give way to you. I passed several in the open plain within fifty paces, they viewed me for a moment as something novel and then very unconcernedly continued to feed."

They saw dens spilling over with young wolves. They found brown bears gorging upon the carcasses of bison. They roasted fish over fires and sliced their bellies open to reveal pale flesh rippling with fat. They ate fine veal and beef, and venison and beaver tails. These were days of plenty; the world was an open casket of delicacies and delights to sample. They shot down more fresh meat than they could ever hope to eat.

It couldn't last. Upstream, game began to grow increasingly scarce—and what deer and antelope they did see were wary and hard to track. Finally, on August 10, 1805, they stumbled across an old Indian road, and later a Native American man on horseback, who fled their advances. This was the first human they had encountered in four months.

Finally, they made contact with the Shoshone tribe who lived in this region, who welcomed them warmly. But the Shoshone were near starvation themselves, and had little in the way of food to offer—a few dried chokecherries, a few mouthfuls of boiled antelope. Game was hard to come by. Lewis watched twenty hunters on horseback pursuing ten antelopes for hours; "about 1 A.M. the hunters returned had not killed a single Antelope, and their horses foaming with sweat." Later, in the densely populated Columbia Basin, they would be forced to eat eleven horses and nearly two hundred dogs.

In entering Shoshone territory, the Corps of Discovery had—perhaps unknowingly—left a vast disputed territory, a debatable land of around forty-six thousand square miles separating at least eight warring tribes. Such regions were an established feature of intertribal relations in America before European settlement; buffer zones of an unmarked kind, well known to the inhabitants of neighboring nations. Hunters would not dare to trespass inside these lawless no-go zones. Only war parties, moving fast. As a result, numbers of prey animals would rebound hugely in the absence of hunting pressure.

In Wisconsin, the Chippewa and Sioux were almost continually at war between 1750 and 1850, forming a buffer zone of up to thirty-eight thousand square miles. In a bitter quirk of nature, it was the reserve-like quality of the zone itself that fed their warfare: in the absence of hunting parties, the wildlife inside recovered to such an extent that the tribes—now wealthy and comfortable—could afford to be magnanimous. Once a treaty was agreed, hunting resumed, deer numbers crashed, and a war over resources recommenced as famine gripped both nations.

These invisible territorial demarcations continue to be a feature of traditional tribal societies today, in the Amazon Basin and Papua New Guinea, among others—also, and not coincidentally, some of the world's richest and most valuable habitats. As Jeffrey McNeely, former chief scientist at the International Union for Conservation of Nature (IUCN), has noted, the buffer zones between warring, pre-state societies serve as refugia for wild game and thus have "helped contribute to the rich biodiversity found today in many tropical forests." Fear, therefore, is a force that shapes the world.

On their way home, toward the end of their great expedition, the Discovery Corps retraced the Missouri River and once more entered into the wonderland experienced the year before. In South Dakota, Clark climbed a hill to gain a lookout of the land ahead and found himself confronted with a vast plain teeming with bison—twenty thousand, maybe more, kicking up dust as they swirled in an ever-moving mass—more bison in one place, wrote Clark, than he had seen in his life. "I have observed," he added, "that in the country between the nations which are at war with each other the greatest numbers of wild animals are to be found."

Yiannakis and I cross the border at checkpoint nine and drive for what feels like miles through a deserted suburb. The road is lined with chain-link fence on both sides, and topped with a coil of razor wire. A thin screen of black sackcloth has been strung up over it, as if that could prevent us from noticing the devastation beyond. It is a ghost town: hundreds of villas spilling off in all directions, in various states of disintegration.

At some stage, all the doors had been removed—to stop the buildings being used as holdouts during the fighting. Windows have been smashed from their frames, shards bristling at every edge. Here and there the cloth is tattered and flapping, revealing thin-ridged succulents crowding on balustrade terraces, the tendrils of jasmine swaying in the breeze, the prickly pear that pushes its fruits through the fence as though selling its wares. Last year's harvest lies dead and dying in the gutters. In the oldest houses, rats have burrowed holes through the red packed-dirt walls. Everything else is crumbling away. I peer through a gap in the cloth. A misted street name: *Derynia Road*.

Elsewhere, the sackcloth is unnecessary—vegetation has risen up to claim the fence, clutching at it and buckling it under its collective weight. It conveys a sense of submersion, of drowning, of being overcome by time. In an unfinished apartment block, makeshift brick battlements have been cobbled together, untidily mortared in uneven rows. Former holdouts, like empty nests, are in evidence on upper balconies, hardboard sheets pulled across open doorways. There is an uneasy sense of war still in progress, as if any second one might take up weapons and run upstairs.

"Put your camera away," snaps Yiannakis, out of character, as a Turkish army vehicle drives past. We have passed now from buffer zone to occupied land. Away to our right rises the skeletal skyline of the abandoned high rise hotels of Varosha—a beachside suburb of Famagusta, once a celebrity haunt frequented by Brigitte Bardot and Richard Burton and Elizabeth Taylor, and now a cordoned-off forbidden zone held by the Turkish army. It was a place to be seen, to be photographed. Now all photographs are forbidden, as a blood-red sign warns in five languages, under the black silhouette of a soldier bearing a rifle. I slide my phone into my pocket, close my notepad, and make a false, stupid pretense of bored disinterest until the patrol passes.

In China Miéville's modern classic of speculative fiction *The City and the City*, two police officers attempt to solve a murder in a Besźel, a city that co-occupies the same site as its "twin" city Ul Qoma. The two cultures these cities represent share a common root and a profound enmity. They communicate in different languages, their citizens living alongside one another, sometimes passing right by one another in "cross-hatched" streets where the territory overlaps, and all of them willfully "unseeing" all that goes on in the rival city.

Miéville's conjoined cities are a work of fiction, but they are what I think of as we arrive in what the Greek Cypriots call

Famagusta and the Turks call Gazimaǧusa, a place where living and dead cities share a site, and the division between them just as equally, vividly, unreal.

On one side, I find a bustling, attractive, medieval city. On the other, a haunted district of crumbling modernist monoliths, pocked and cratered with shell scars. All that separates it from us is a flimsy wire fence strung with sackcloth, which everywhere is rucked and torn or flutters loose. Beyond it I see streets knee-deep in grass and golden flowers. Broad palm fronds throw arms over the fence, as if hailing cabs. A stage set for a horror movie, and yet, all around, life rushes incuriously by, as if one might simply *unsee* the chaos, the bad twin beyond.

The abandoned skyrise hotels of Varosha are best viewed from the beach. A shelf of silver sand extends a short distance into a sea of green. Close to the border, such that it is, a few ludicrous attempts at educating tourists in the art of unseeing have been annotated: the usual blood-red signs, and NO PHOTO scrawled in black paint like graffiti across a concrete balcony, below which black cloth blows coyly from its anchors like a bridal veil.

A barricade has been thrown up where the Varosha frontier crosses the beach: a hodgepodge of corrugated sheeting of various lengths, long-stained and poxed with oxidation. Several oil drums weighed down with cement list together at angles. I could climb over it in a second. I could wade knee-deep into the surf and walk around it. In *The City and the City*, those who interact with the enemy state overlaid upon their own—*BREACH, BREACH*—are spirited away by the all-seeing secret police. Here, all that stops me is the soldier in the nearby observation post, a spotty teenager toting an oversized gun who watches me: a lone woman on a windswept beach, staring off into the forbidden zone.

A few years ago Paul Dobraszczyk stood here in this same spot, considering the very same sights. A British academic in his thirties, he came here for very similar reasons: to stare out at the decaying concrete blocks that rise all along the beach to the south. He stayed at the Palm Beach Hotel, a luxury resort that stands strangely oblivious on the silver scoop of sand at the beach's far north end, as if unaware of the crumbling dystopia on its doorstep.

The following day, he began to walk the sackcloth boundary, not with any particular intent, just to follow it around, get a sense of its full extent. No one paid him much attention. But then, after a while, he came unexpectedly upon a gap in the fence.

Heart racing, but without thinking too hard about what he was doing, he stepped forward and through the hole—*BREACH*—and found himself alone in the ghost city. It was, he says, frightening. Like stepping through a portal into a different world. All the commotion of the living city, the cars and the voices and the footsteps, continued on regardless, obscured behind the rumpled sackcloth. Cautious of the grave implications of being caught—arrest and deportation, certainly, and one couldn't discount those rifles on those red signs—Dobraszczyk crept forward into the deserted street. And here he was, the last man on Earth, walking alone through the end of time.

Then, as soon as it had begun, the illusion was shattered. A second man stepped out into the gloom. A looter, probably, or a scrapper of some kind, but Dobraszczyk didn't pause to find out. He shied away, darted back to the fence and back through the portal into the present. It felt like a close call.

Later, though, he found his courage restored. He returned to the hole in the fence, slipped back through. This time he saw no one. He entered open doors and found himself in high-ceilinged rooms where paint flaked like petals onto hardwood floors; in a mall, where trees growing from the shopfronts stretched their thin limbs up toward the skylights; in an apartment block, where he rested a few moments, allowing his pulse to return to normal.

Greek paperwork lay scattered across the desks; cutlery and plates on the tables. From somewhere outside, not far away, he heard the call to prayer in Gazimağusa, and allowed himself to listen, to coexist in both places at once. He felt, he remembers now, an incredible sense of peace. Just being there felt like an immersive kind of meditation.

What struck him was how well preserved the rooms were, given the length of time they had been left empty: the air inside was dry and clean, and the only noise the rustle of the wind and the low voices of the pigeons that had made the apartment their home. Five floors up, on the roof, he took in the view of this whole dead city stretched out before him, almost overwhelming in its scale, in all its sublime unknowability.

On my way back, I pass the military checkpoint for access into this disheveled paradise at John F. Kennedy Avenue. I can see beyond the barrier to the scene of Dobraszczyk's epiphany, past the abandoned gas station, into the collapsing streets grown choked with green, the old advertisements faded and bleached; a city of the 1970s, weathered and weathering the storm.

In the summer of 2008, a group of scientists gathered before dawn—seven Greek Cypriot and seven Turkish Cypriot scientists—at a UN checkpoint, where they were ushered into vehicles and driven into no man's land. They were there to begin a year-long study of the plants and animals that had taken over the ownership of the land in the decades since its abandonment.

They chose eight locations within the buffer zone, cutting across the island as a cross section of Cypriot landscapes, from the sand beaches and flood plains of the coast near Famagusta, through rich wetlands, to the mountains and rocky coast of the far west. Together, these places offered a glimpse into what the island might look like should people, one day, be gone.

The scientists worked fast, setting their camera traps and laying quadrats. Anxieties ran high, as they often worked in full view of the rival watch towers; though they had permission, who knew if the message had got through to the squads on duty. Once a month or more, over the course of a year, they returned to the same sites, building up a picture of what was going on when there was no one there to watch.

One of their sites was the old Nicosia airport, once the scene of a running battle lasting days—tanks and anti-aircraft artillery looming out of the smoke of the brushfires, air thick with the petroleum reek of napalm. There, in the cavernous departure lounge, rows of ergonomic chairs sit piled with guano, dramatically downlit by circular skylights; ceiling tiles peel away in long skin-like strips to reveal dangling wires. Posters for long-discontinued holiday deals slip down in their light boxes; hoary cobwebs mottle the glass. Outside, the wreckage of a jet, crash-landed, lies like a deer being dressed, guts spilling across the rough ground, the flag of former Czechoslovakia on its tail fin. An RAF patrol plane, studded with bullet holes, rests at the end of the runway on a cushion of thorns.

The camera traps revealed life amid the wreckage. As well as the pigeons, barn owls were bedding down in cracks and holes in the masonry. Snakes basked on the cracked runway. Foxes hunted for mice in the long grass. Falcons nested on the roof of the control tower. "These animals are very sensitive toward humans," says Dr. Salih Gücel, the co-leader of the project. On an island as densely populated as Cyprus, where every scrap of land is accounted for, and hunting is extremely popular, any bolthole will do.

Elsewhere in the buffer zone, exceptionally rare plants including the Cyprus bee orchid—with its pouting, velveteen lip whose markings mimic a female bee—and the vanishingly rare Cyprus tulip, with its crimson petals, grow in large numbers. In all, the scientists recorded 358 species of plants, 100 species of birds, 20 reptiles and amphibians, and 18 mammals. The spoils of war.

But of course it is not the *war* that has had a positive impact. War in general is equally destructive for humans and the environment alike. Aggressive tactics saw millions of acres of Vietnamese and Cambodian rainforest sprayed with defoliant in the name of war. Scorched earth policies saw more than a billion barrels of crude oil set alight in Kuwait. The horrific Rwandan civil war sent more than a million refugees fleeing for their lives into Congo's Virunga National Park—home to Dian Fossey's beloved mountain gorillas—where they were forced to kill the animals they found there for food, and fell its trees for fuel and the construction of camps.

But the *exclusion* of people, by way of a no man's land, is a different matter altogether. Stability is the key ingredient: stalemates of the sort seen in Cyprus, or deadly hangovers from wartime such as minefields, create no-go zones not unlike strict reserves, protecting wildlife and halting the exploitation of natural resources. Such results are, of course, merely a happy side effect of an extremely unhappy situation. But they have taught us some valuable lessons.

The Second World War, for example, inadvertently tested and proved the concept of the marine protected area: for six years, fishing in British waters effectively came to a halt. Fishing vessels were sunk or requisitioned—creating a de facto marine reserve in the

North Sea of around 220,000 square miles. During this relatively brief interlude, wild fish stock rebounded—and catches shot up dramatically when they resumed fishing after the war, in bounteous harvests that lasted for around a decade before they were again depleted.

The Iran–Iraq war between 1980 and 1988 saw miles of borderland between the two countries planted with upward of twenty million land mines, rendering it a dangerous no-go zone. More than thirteen thousand people have been killed or injured by explosive devices in Kurdistan alone in the years since. But now the region has become the most significant stronghold of the endangered Persian leopard, thought to number fewer than a thousand individuals in the wild. Although the big cats weigh up to one hundred and eighty pounds, they rarely put all their weight on one paw, and thus escape death by the Soviet-era munitions. Similarly, Yorke Bay—once a popular beach near Stanley, the capital of the Falkland Islands—became, effectively, a strict nature reserve after mines were laid there by Argentine forces during the 1982 conflict, and is now home to a bustling colony of Magellanic and Gentoo penguins.

As we saw in the case of the Colorado bison, it can be that such benefits last only as long as the wars themselves. (Indeed, the beach at Yorke Bay is currently being cleared of mines in line with the Ottawa Convention, despite local opposition and at great disruption to the ecosystem.) But the unplanned nature preserves that have formed up in buffer zones have come to serve as a focus for bilateral cooperation after hostilities are over.

During the Cold War, for example, the heavily guarded Inner German border—stretching from the Baltic Sea to the border with Czechoslovakia—became a favored haunt of birders on both sides. Though the "death strip" itself was plowed bare and spotlit (not to mention tripwired, booby-trapped, and patrolled by East German soldiers with a shoot-to-kill policy), a restricted area of between one hundred feet and several miles in width was maintained by the East Germans along the entire inner perimeter of the border, and thus given a reprieve from the otherwise intensive agriculture in the region. Black storks, nightjars, and red-backed shrikes nested in the branches between control towers. Lady's slipper orchids blossomed along the forest edge. Moor frogs spawned and otters paddled in the anti-vehicle ditches. Ecologists often refer to "wildlife corridors," strips of wild land that serve as links between habitats: here was a green highway 860 miles long, offering safe passage through an entire country. Over the forty-five years of its existence, this contested hinterland—much of which had previously been valuable farmland—was colonized by more than a thousand species from Germany's "red list" of endangered species.

During German reunification, three hundred amateur ornithologists from both sides of the border organized an emergency meeting in a tavern, where they hammered out a manifesto for conserving the death strip as a nature reserve. Their success went on to inspire the wider "European Green Belt" movement, which now takes the form of forty linked reserves in twenty-four countries that fall along the course of the former Iron Curtain, taking in dense boreal forest along the Finland–Russia border; the sand dunes, cliffs, and lagoon of the Baltic coast; and a belt of mountainous uplands through the Balkans, where lynx roam and imperial eagles soar overhead.

Surely the most valuable environmental outcome of the Cold War, however, must be the Antarctic Treaty—negotiated during a brief thawing of hostilities in 1959—during which time the countries that had staked, or planned to stake, territorial claims during the "scramble for Antarctica" agreed to set these claims aside so as to create "... a natural reserve, devoted to peace and science." The treaty will come up for review in 2048.

And more recently, the environmental value of buffer zones has come to be recognized as valuable elements in the peacebuilding process itself. For more than a hundred and fifty years, Peru and Ecuador were locked in a bitter territorial dispute over the Cordillera del Cóndor, an offshoot of the Andes rising between the two countries, leaving large areas undeveloped as a result: pristine forests unlogged and rich gold and copper seams unmined. Environmental surveys in the 1990s revealed the region to be one of the most biologically diverse (and least-studied) habitats in the world; almost every visit to its slopes reveals yet more species unknown to science. This environmental storehouse became a key plank of talks—something of significance that the two countries now shared—and, as part of the 1998 peace agreement, both sides committed to creating extensive reserves on both sides of the border. Transnational reserves of this sort are known as "peace parks"—powerful demonstrations of the healing power of nature, in more ways than one.

One might hope that such a convivial agreement might be reached over the demilitarized zone (DMZ) between North and South Korea, where soldiers face off across a 155-mile-long, 2.5-mile-wide strip of no man's land. This zone, plus (to a lesser extent) the neighboring strip behind the "civilian control line" on the south side, has been walled off and tightly policed since 1953. To the urbanized south, the land has been dug up and developed; to the poverty-stricken north, much of the forest has been shorn for use as firewood. But the strip of temperate forest, wetlands, and abandoned paddy fields caged within is home to thousands of species otherwise extinct or endangered on the Korean peninsula. The Asiatic black bear, the Korean water deer, the rare long-tailed goral, and the diminutive leopard cat have all been spotted thriving in the DMZ, amid the land mines and tank traps. Around twenty thousand migratory birds use the border zone as a stopping-off area each year; it also hosts the world's largest population of endangered red-crowned cranes (elegant creatures that perform balletic courtship duets, considered in Korea a symbol of peace). Sightings of the Amur leopard and Siberian tiger—two of the most endangered cats in the world—have also been reported. One South Korean soldier recalled his time stationed in the DMZ as like "a natural paradise," where he saw more wildlife than ever before in his life. But his nights there were punctuated with the bass drum of explosions, as the animals themselves set off mines and tripwires.

South Korea began to pursue the idea of turning the Korean DMZ into a nature reserve in 2011, and suggested in a 2014 statement that such a park could come to form a symbol of peace between the two countries. It has persisted with the idea, laying the groundwork along its own side of the border, and in June 2019, successfully applied to have nine hundred and sixty-five square miles of land falling in this "civilian control zone" designated as a "biosphere reserve" by UNESCO.

So far, North Korea has declined to engage.

The day before I leave Cyprus, I hire a car and drive inland, away from the gaunt towers of Varosha, hugging the buffer zone as tightly as I can.

Boundaries are complex on the east of the island. I travel for a while on a road where either shoulder lies in control of opposite sides. I pass a dead-eyed village of houses and a church, stripped to their sand-colored shells. This is Achna, whose residents were displaced, and is now in use by the Turkish military for urban environment training—and thus lives and relives the worst day of its existence again and again.

In Nicosia, the capital, the buffer zone takes the form of a single narrow street that winds through the very heart of the city, splitting it in two. I wander the old city, along streets that dead-end unexpectedly in oil-drum blockades painted in flaking patriotic paint. Beyond I see shop facades, faded and dark with mold, their roofs and sills flush with vegetation. Barricades of sandbags sag on balconies. Here and there the ornate stonework is marred by gunfire and mortars. Where it is most impacted, it takes on an esthetic all its own: rough-chiseled like a sculpture partway through carving. Or else acid-splashed, half-dissolved, semi-liquid.

On a whim, I cross the border on Ledra Road, lining up to show my passport, get it stamped a few yards later. More signage in the interzone, demanding NO PHOTOS. High fences are thrown up, as if to stop even curious eyes. I think again of *The City and the City*, feel my focus switch to the Turkish signs, the minaret. I drink a coffee, then cross back. Legally speaking, of course, I have never left: the existence of the Turkish Republic of North Cyprus is recognized by no country other than Turkey.

I head west, past the old airport, and weave up into steep, remote country. Pointillist hills are stippled with shrubbery. The rock is orange-gray in places, blue-gray in others. This is the most remote region of the island, where the buffer zone is widest. Observation towers here stand far apart, and sometimes abandoned. I pass more than one—painted in amateurish camouflage of sand and dun—where sun-bleached signs warn against ingress, their alarming tone somewhat undone by the golden floral halo pillowing up around their legs.

Finally, I see it. I pull up close to a UN helicopter pad that is baking in the sun, and take out my binoculars. Across a valley of no man's land, a second mountain arises between me and the sea. On the slope facing me, I see the village of Variseia.

Even from far away, it's clear that this is no ordinary village. It has a ghostly presence; even looking at it gives me a chill. There are gaps where the windows and doors should be. Pale gray walls crumble into dust. Red roofs peek between trees that press tightly in against the whitewashed walls, as if supporting them through their collapse. The terraced landscape is softened—eroded and overgrown. But though the village is long abandoned, I see movement. Birds flit from the windows and roofs.

It was here that Salih Gücel and his colleagues found their most striking results of all. In this empty village, high in the hills, they discovered the highest species richness of all the sites they surveyed. Cyprus mouflon—a wild sheep found only on this island—were discovered to have reoccupied the empty houses, using them for shelter from the sun and winter storms. Numbers of these hardy dwarf sheep, with their russet coats and huge scythe-like horns, fell to only a few dozen in the twentieth century, despite their cherished status as a national symbol, yet here they were, dozens of them: caught on film, looking off-camera, and moving with a relaxed stride. During the stalemate the mouflon have thrived in the abandoned villages, roaming in large groups down empty lanes, grazing the farmland left fallow. In all, through a combination of huge areas of new habitat in the buffer zone, and stricter protections, numbers are thought to have rebounded to three thousand or more.

Also recorded slipping through the deserted village: barn owls with their pale, heart-shaped faces hunting the rare endemic mouse, *Mus cypriaca*; brown hares loping wide-eyed through the long grass; blunt-nosed vipers, spiny lizards, and the leopard-spotted camouflage of the Mediterranean chameleon.

Here, in the silent valley, where golden pollen dusts my skin and birdsong drifts through the air, war feels a long way off. The sun beats warm through a thin haze of cloud. A sea breeze ruffles the trees. A cicada rasps a scale, first up and then down. Songbirds tumble in the air, oblivious to my presence: sparrows and wheatears, with black back and necks, and rose-taupe bellies. A swallow strafes the road, swerving this way, then that.

OLD FIELDS

Harju, Estonia

The greenhouse stands chest-deep in thistles. They grow tightly packed, straight up, soft heads fluffed and overripe, coming loose at the sides, ruffled by the drafts that seep through the broken panes. Thistledown hangs in the air, shifting on almost imperceptible currents, moving slow through shafts of light.

Drawing my hands inside my sleeves, I push politely between the stems as if through a crowd. I come into a thicket of raspberries— survivors from a more cultivated era—whose thorned limbs have grown into a dense and tangled knot, their fruit clenched like tiny knuckles of sage and dusky pink, rot-black. Tiny insects drift above like ash in smoke, and over that, a trellis of rusted pipes that once were arteries, pumping hot water through this great, glazed hall in its glory days.

This is rural Estonia, the former site of a collective farm, a kolkhoz. Once a bustling center of local agriculture, the greenhouses—here, and hundreds like them—have fallen silent and disused, the old crops dead or growing feral. In the former Soviet Union, kolkhozy like this were once the norm: the result of a Stalinist policy that sought to boost production and free the peasant class from servitude. Lofty goals; in practice, collectivization dispossessed millions of their land, sparked civil disorder, and, at times, produced food shortages and famine.

Most of Estonia's farmland was forced into collective ownership during a single turbulent month in 1949, when eighty thousand resisting kulaks (landowners) were deported to Siberia and Kazakhstan. Those who remained surrendered their fields and their animals to the common pot, and—as one contemporary Communist Party report put it—"quietened down and concentrated on intense everyday work."

In this way, traditional family farms were put aside in favor of monolithic enterprises. Brutalist Khrushchyovka apartment blocks, five stories high, appeared incongruously on the edges of tiny hamlets, or next to agricultural buildings of industrial proportions. Sheds like aircraft hangars; grain elevators like oil terminals; barns built to hold ten thousand head of cattle. Grand manor houses, dating from the tsarist era, were emptied of furniture and inhabitants, repurposed as administrative hubs. It was the new way of farming, the new way of working, the new way of living.

It didn't last. When, in 1991, the Soviet Union disintegrated, it precipitated one of the biggest revolutions in land use the world has ever seen. All across the former USSR, collective farms—heavily regulated and state-subsidized—were forced almost overnight to weather the vagaries of the free market. Since then, nearly a third of all Soviet farmland—an estimated 245 million square miles, an area roughly the size of France—has been abandoned.

In Estonia, in a symbol of their rejection of communism, the kolkhozy were subdivided and redistributed among the previous owners and their heirs—many of whom had since left for the city, or emigrated abroad, or simply did not want to farm. Land fell into disuse. The huge depots and warehouses, too large for civilian use, stand empty, monuments to the past regime. Like here, where I stand, in the greenhouse.

A jungle of vines hangs across the glass hall like a blackout curtain. Long past their best, they have bloomed and fruited unattended, throwing down their grapes in Grecian splendor only to see them wither on the vine. The grapes are small, soft, sunken. Sickly green and brown, furred in places, a lure only for the birds, who squeeze in through the shards to find themselves trapped inside a tunnel of light.

Outside, the wind is picking up. A branch taps feebly against the glass. It's getting dark.

Tarmo Pilving was eleven when the Soviet Union fell. The early nineties were a bad time for Estonia. For a time its capital ranked among the world's most violent cities, as the old rules fell away and a new class of mobsters and would-be oligarchs wrestled for control. But, for a teenage boy, it had its benefits.

The departure of the Soviet administration (and, later, the Russian army) left a great many buildings and compounds empty and unguarded, ripe for exploration and ready to serve as the backdrop to a generation's coming of age. Tarmo passed a misspent youth running wild through empty hallways, climbing dark staircases, drinking scavenged beer, and playing football in overgrown fields.

One favored haunt, only a short bike ride from his home near Tallinn, was the former missile base at Türisalu, where, until recently, nuclear warheads had waited, poised for action, pointed to the west. When the army pulled out, they did so without warning and overnight, leaving the shells of buildings and huge, grassed-over landworks, and the gates swung open. Once Tarmo and his friends stumbled across an anti-tank mine, forgotten in the grass since the occupation. They tried to detonate it by throwing it on a campfire. Sometimes, he says, he thinks he's lucky to be alive.

Tarmo grew into a thoughtful, sandy-haired man with a fluting voice, now a researcher at Tartu's University of Life Sciences. As he has grown and matured, so has his country. And as Estonia has grown increasingly wealthy and increasingly urban, so too has the land around him.

Not far from the old base, we stop at a lightly wooded grassland a hundred or so yards inland from the Baltic coast. Tarmo's childhood home was just around the corner; he recalls standing in this same spot as a child, and the whole expanse ahead a rippling sea of green barley, ripening slowly in the sun. He would wade into it up to his hips, imagine it washing over him, see the swell spooling out across the bay.

When the state-run farm that tilled this land, Ranna Sovkhoz, was dismantled, the field—like many others—fell into disuse. Within a year, it had shed its skin: the barley, dependent upon human intervention to survive, did not return. Since then, a changing cast of species has paraded through, as part of a process so fast as to make unrecognizable entire landscapes within a few short years, yet so slow as to be undetectable to the human eye. The plants shift in their seats, trade places, multiply, and disappear—but only when we are not looking. To stand in the field and consider its progress is to confront the eerie sensation of having been elected the unwitting judge of a game of musical statues; the trees and plants frozen in comic poses, their riffling leaves giving them as effectively away as the shallow breath of the stilled but living body.

First came the wildflowers, the annuals, the weeds. Later the thorn bushes, the brambles. Now, the old field appears in a state of dishabille. A ragtag crowd of skinny saplings of various shapes and sizes mill around, waiting for something to happen. Bare rowans bearing handfuls of bright-flashing berries; slim and silk-skinned birches and aspens, their leaves aquiver; a thicket of thin and whippy willows, tightly spaced as cigarettes in a pack. Here and there stand the hunched backs of juniper, whose dry and aromatic boughs predominate in these deserted grasslands. There are a few bushy shrubs—dog roses, long-faded but lit with festive rosehips, trail out along the edgelands—but mainly there were the thin, spare grasses that gave the pleasant aspect of a wooded meadow.

Tarmo beckons me over to a flattened area of the grass about the size of a double bed, where a large animal, probably a moose, had lain down to sleep and I feel a shiver of something like unease, from the sense of large animals being so close at hand—even here, only a few miles from the capital.

He parts the grass to reveal sproutlings of spruce. Only a couple of years old, these scraggy toilet-brush trees look wobbly and vulnerable, but soon, says Tarmo, all this will be theirs. In Estonia, after a certain period of time, abandoned land will almost always come to take its final form as a dark, dense spruce forest. These toddlers' slow but continual rise will, at some stage, enshadow and ultimately overpower the pretty broadleafs whose gilt and rubescent foliage shimmers all around.

The metamorphosis of abandoned fields like this one is the classical example of a concept that forms the very heart of ecology. "Succession"—the process by which bare ground may come, in time, to transmute into forest—has been as central to the field as evolution is now to general biology.

You probably studied it in high school: the classical model, based on the thinking of the Nebraskan botanist Frederic Clements, goes something like this: a plowed field, left to go fallow, will over time pass through a number of intermediate stages—"seres"—as the age of the weedy annual passes into the era of the shrub, and then the kingdom of the fast-growing softwood trees, before finally, over a period of many years, the process climaxes in an established hardwood forest, which remains in place until the land should be disturbed again.

And though the specifics have since been disputed (most notably, the concept of a stable "climax" has been discarded in favor of a more dynamic final stage, with a shifting cast of species according to climate), the swaths of former farmland here in Estonia are living proof of this natural law—the tendency for abandoned land to turn, over time, into forest.

As a result, tree cover in Estonia has been fairly constantly and quite rapidly increasing: from only 21 percent of the country in 1920 to 54 percent of the country by 2010, in all gaining around two thousand square miles of forest since the fall of the Soviet Union. Estonia is now one of the most forested countries in Europe—and 90 percent of that forest has "naturally regenerated."

The same pattern has been borne out across the former Soviet Union. One 2015 analysis of satellite images estimated at least forty thousand square miles of forest regrowth in eastern Europe and European Russia alone—noting that only an estimated 14 percent of the abandoned farmland had yet converted, thus raising the prospect of large-scale carbon sequestration well into the future.

An unexpected outcome, then, of the collapse of the Soviet Union: the biggest man-made carbon sink in history.

A 2019 study attempted to quantify the impact in those terms, suggesting that between 1992 and 2011 the carbon sequestration in the soil of the abandoned farmland, combined with the decline in meat and milk production in the wake of the political upheaval, is equivalent to a reduction in carbon dioxide emissions of 7.6 gigatons. This is around a quarter of the emissions generated due to deforestation in Latin America over the same period, and is, the researchers were keen to underline, likely a "substantial under-estimation" as they have not yet taken into account the carbon held in the vegetation itself.

Other scientists had previously attempted a similar feat. Estimates vary considerably according to methodology, but one 2013 study estimated the carbon sink in Russian territory only at 42.5 million tons of carbon a year, every year since 1990—equivalent to 10 percent of Russia's emissions from fossil fuels. If these figures are correct, this would mean that Russia has, technically, easily surpassed the terms of the Kyoto Protocol through the abandonment of farmland alone. (The study added that another 261 million tons could be sequestered over the next thirty years.)

But the point is not that land abandonment, on a large enough scale, would permit us to go on burning fossil fuels without consequence. There is not enough land in the world to allow us to keep digging carbon from its grave—where it has slept for a hundred million years—and setting it loose. What forest regrowth offers us is a chance to pay our debts, to atone for past sins. It is not a pardon, but a reprieve.

It may too be the answer to a question that has been confounding scientists for decades.

Those who work in the field of Earth systems have spent many years attempting to balance the global carbon budget. To do this, they calculate the total carbon emissions generated from the burning of fossil fuels, and attempt to match that figure to the carbon known to be held in the atmosphere, on the land, and in the ocean. But the sums don't work—essentially, the carbon levels of the atmosphere have not been rising as quickly as we expected, and almost certainly because carbon is being sequestered on an enormous scale *somewhere* in the terrestrial or oceanic spheres.

Various theories have been proposed since the problem was identified in the early 1990s—recent suggestions include: the carbon is stored in massive aquifers under the ground; that carbon is sitting at the bottom of endorheic basins in desert lands; and that the rising carbon dioxide levels of the atmosphere and attendant rise in global temperatures have stimulated plant growth, and thus elevated carbon storage—a negative feedback loop of the sort proposed in James Lovelock and Lynn Margulis's famous Gaia hypothesis, in which the biosphere is essentially proposed to function as a complex, self-regulating superorganism.

Lovelock and Margulis named their theory for the Greek goddess of the Earth, and though it is a powerful metaphor, they do not mean to suggest that the planet is imbued with deific powers. Still, it can be hard for those of a poetic disposition not to interpret the apparent drag on atmospheric carbon levels as anything other than divine providence: an act of forgiveness on a planetary scale, or of self-sacrifice, as the Earth shields us bodily from our worst excesses.

But the authors of the 2019 study believe that abandoned Soviet farmland might account for at least "a considerable part" of this missing carbon. Certainly, much of it is in the optimal climatic zone for sequestration: recent research suggests that young forests growing in temperate climates absorb and fix carbon at higher rates than in the tropics or subarctic.

To find a solution to the case of the missing carbon, but for that answer to be enormous political and social upheaval, feels—to say the least—a mixed blessing. Perhaps what we could take from it is that the tools for massive carbon sequestration are already within our grasp. As the geographer Susanna Hecht has said: "Trees have already been invented." All that the Earth may need to soak up enormous, climate-altering quantities of carbon is to be left alone.

The former Soviet Union is far from the only region in the world to have seen significant regrowth in recent decades. Over the last century, there has been a "dramatic and ongoing" increase in the amount of abandoned farmland worldwide.

Quantifying this land is notoriously difficult, given that abandonment often goes unrecorded, but a global analysis in 1999 found that abandoned land began to rise steeply around the turn of the twentieth century, and quickly accelerated toward the end, with the United States, China, South America, and the former Soviet Union making the largest contributions in real terms.

In the United States, agricultural abandonment on a large scale began as far back as the 1860s, as farmers left the rocky, acidic soils of New England in favor of converting those boundless flat expanses of the Midwest into America's "breadbasket." I myself have walked in the woods of New England, through dense and apparently untouched forest, only to stumble across a tumbledown stone wall, furred with lichens and moss, marking the boundary of what was once a carefully tended field. These were the original "old fields" that served as research fodder for much early work on succession theory, building upon Clements's seral vision.

It is strange to think now, but when Henry David Thoreau

was living his "Life in the Woods" at Walden, deforestation in the northeastern United States was, in fact, at its peak: only 10.5 percent of Concord's woodlands were standing by 1850. Thoreau's own cabin was built in a recently logged clearing, and Thoreau himself found work as a surveyor of the neighboring lots, parceling up the woodland ready for harvesting.

But, as Thoreau knew: "One generation abandons the enterprises of another like stranded vessels." Agricultural land in the northeastern United States nosedived from 72,500 square miles in 1880 to 18,800 square miles in 1997, and as it did, the forest was quick to reclaim it. New England is the most thickly wooded

Changes in forest distribution over 400 years.

region of America, from around 30 percent cover then to over 80 percent now. And with it, numbers of beaver and moose and white-tailed deer and bears and woodpeckers have rebounded too. If wildness truly is the preservation of the world, then surely New England has been moving in the right direction. Certainly, today's Walden Woods more closely resembles the wild forest of the book than it did in 1845. Abandonment would go on to spread through the eastern seaboard, and later through the American South. All in all, the American forest grew by around 1,400 square miles every year between 1910 and 1979.

More recently, rural abandonment has become a significant trend in China, Latin America, and Europe. In the European Union alone, an area roughly the size of Italy is expected to be abandoned between 2000 and 2030. In rural Spain, for example—most notably in picturesque Galicia—so many have been trading their ancestral homes for the city that an estimated three thousand ghost villages, in various stages of dereliction, lie empty. Spain has tripled its forest area since 1900. And abandoned farmland has been a significant factor in the startling and largely self-directed return of large carnivores to western Europe: populations of lynx, wolverines, and brown bears have spiked. In Spain, the Iberian wolf has rebounded from four hundred individuals to more than two thousand, mostly to be found haunting the empty villages of Castilla y León and Galicia and feasting upon the wild boar and roe deer, whose numbers too have ballooned. In early 2020, a brown bear was spotted in Galicia for the first time in a hundred and fifty years.

Worldwide, though deforestation remains a serious, pressing issue in the tropics, a major survey based on thirty-five years of satellite imagery, and published in *Nature* in 2018, contradicted long-held assumptions of decline by suggesting that global forest cover has actually grown by around 7 percent, or roughly 860,000 square miles, since 1982. Not all of this is due to abandonment—industrial plantations also count toward the total—but so marked has the recent reversal in the fortunes of the forest, in so many disparate regions, that geographers have started to describe a country displaying large-scale reforestation as having passed through its "forest transition."

Like Clements's orderly model of succession before it, the forest transition concept brings a reassuring sense of determinism. It suggests that every country might pass through its phase of environmental destruction like a difficult adolescence, and then rebound: dense forest to flayed, denuded landscape, and a transformation back into a land of plenty. Currently, forests are declining in around a third of the world's countries, stable in a third, and growing in the final third. How comforting to imagine that it is simply a case of waiting for two-thirds of the world to catch up.

And, well—maybe. Though forests in the tropics remain in retreat,* this rate is slowing. Thirteen tropical countries are considered to have graduated through, or are approaching, forest transition. In a 2011 discussion of forest transition, the environmental

---

*By around twenty-one thousand square miles a year.

scientists Patrick Meyfroidt and Eric Lambin raised an exciting prospect: that we could see in our lifetimes a reversal of deforestation. The new *Nature* study raised one even more thrilling: we may already have.

Even in the Amazon, upon which much of our current deforestation anxiety is, rightly, focused, deforestation rates have fallen steeply (by around 80 percent) over recent decades—although a recent surge of clearance under relaxed environmental policies in Jair Bolsonaro's Brazil threatens to reverse this trend—and large tracts of previously cleared rainforest are deserted each year. These tracts are usually degraded by soil erosion, and compacted by grazing cattle. The secondary rainforest regenerating there cannot yet claim even a fraction of the richness of what was there before, and it will be many, many decades, even centuries, before it can. But it's a start.

Overall, more than two-thirds of the world's forest is now considered "naturally regenerated." This is Christ-like rebirth, Lazarus-like revival, on land we had left for dead.

Though the idea of "forest transition" as an economic stage is a new one, large-scale abandonment of cultivated land is not unprecedented. There have been periods of history—cursed periods, periods wracked by war, or periods struck down with plagues of apocalyptic proportions—when farmland has been abandoned wholesale, for decades or even centuries. Through the study of such periods, however distant, we might—with difficulty—begin to trace lines between the dots of abandonment, carbon sequestration, and climate.

Starting in the thirteenth century, the Mongol conquests saw somewhere between an estimated twenty million to forty million people die as Genghis Khan and his descendants battled over, massacred, laid siege to, and otherwise rampaged through China, Central Asia, and, ultimately, eastern Europe. The mounted Mongol armies were swift and merciless, laying entire kingdoms to waste and subjecting pockets of resistance to untold ultraviolence. (One thirteenth-century diplomat reported approaching a "hill of snow" only to discover it to comprise entirely of "the bones of men slain" during the sack of Zhongdu;* the road was "greasy and dark from human fat"—a bloodstain so large it took three days to traverse.)

Plagues and famines followed in their wake; centuries-old irrigation systems were destroyed by the hordes in Mesopotamia and Afghanistan. Some estimates suggest regions of northern China were depopulated by as much as 86 percent, as millions fled to the south. In a 2011 paper, an international team led by Julia Pongratz of the Carnegie Institution's Department of Global Ecology attempted to quantify the impact of these events. They estimated the resulting abandoned land at 120,000 square miles, and—hazarding that around half of that regenerated as forest—calculated that the regrowth might have absorbed the equivalent of 700 million tons

---

*Today's Beijing.

of carbon dioxide from the atmosphere, enough to cause a drop in atmospheric carbon.*

Not long after the conquests, a second disaster befell Eurasia, also in a way linked to the Mongols: the Black Death. Sometime in the early fourteenth century, a mysterious illness began to sweep central Asia. At first a patient would come down with what seemed like a flu. Soon festering, boil-like lumps would blow up under their skin—lymph nodes swollen to the size and shape of plums—then rupture like rotten fruit. Sufferers, delirious with fever and consumed by thirst, cast themselves into fountains and public cisterns. Blood streamed from their noses, spewed from their mouths in jets.

By 1347, the busy port of Caffa, in the Crimea, had been under siege by the Mongol horde for months. This was a well-worn tactic of the Mongols, but this time, outside the city walls, a plague was taking hold. "Behold," wrote the notary Gabriele de' Mussi, "the whole army was affected by a disease which . . . killed thousands upon thousands every day. It was as though arrows were raining down from heaven."

Facing defeat, the Mongol army took a most terrible vengeance: "they ordered corpses to be placed in catapults . . . What seemed like mountains of dead were thrown into the city, and the Christians could not hide or flee or escape from them." The plague had breached the walls, and soon the city was in chaos.

---

*This figure is still modest relative to the scale of today's emissions, at approximately the amount produced annually through global gas consumption. Other researchers (e.g., Simon L. Lewis and Mark A. Maslin, "Defining the Anthropocene," *Nature*, March 12, 2015) contend that the Pongratz et al. estimates are overconservative.

Genoese and Venetian merchants fled by sea, skipping desperately along the coast, spreading pestilence wherever they touched down for supplies. In Messina, Sicily, those on the docks were horrified to receive twelve galleys full of sailors either dead or dying, bruised and lumpen with oozing sores, their fingers, toes, and noses black with necrosis. These ships of death, and others like them, carried the plague to Corsica, Genoa, Marseille, Venice—and from then on into Pisa, Florence, Rome, Paris, London . . .

The Black Death swept the continent as an unremitting wave moving north at around 2.5 miles a day. Within three years it washed up in the Arctic Circle, killing three of every five of those it infected on the way. In a few, short, unimaginably horrifying years, an estimated 40 percent of the population of Europe had been killed. Between one-fifth and a quarter of all settlements were abandoned. Harvests went unreaped. Crops rotted in the fields. The abbot of St. Martin's, in Tournay, France, wrote in 1349 of "cattle wandering without herdsmen in fields, towns, and waste lands . . . barns and wine-cellars standing wide open, houses empty." All across the continent, he wrote, fields were lying uncultivated, the farmers and laborers who had tended them all having died. A desolate scene, and one that persisted for many decades to come.

In 1440, nearly a century later, the bishop of Lisieux saw a vast deserted land stretching from the Loire to the Somme, all "overgrown with brambles and bushes." Succession had been set in motion; scrubland and woodland reclaiming old territory. In this way, a forest might reclaim an abandoned field in fifty years or so. Many of Germany's forests today date from this period.

Pongratz's approach, in weighing the carbon sink of disaster, was building upon the work of the American paleoclimatologist William Ruddiman, who in 2003 proposed that human activity may have begun to affect the world's climate thousands of years earlier than previously supposed. In that paper, Ruddiman hypothesized that human activity would have begun to impact the atmosphere as far back as eight thousand years ago, from the dawning of the age of agriculture (rather than the Industrial Revolution onward, as is generally accepted). And so, periods of human die-back—to put it in the bluntest terms—and the decline in farming, should be observable in the atmospheric records.

And it was the Black Death that first attracted Ruddiman's attention. What he found interesting was the timing. Analysis of Antarctic ice cores (which store tiny bubbles trapped for thousands of years in an archive made of ice) demonstrate significant and unexplained drops in atmospheric carbon dioxide (an anomaly of around -5 to -10 parts per million) around the time of the Black Death pandemic. Ruddiman calculated that the drop could be accounted for, had 14 to 27 gigatons of carbon been extracted from the atmosphere and sequestered in new forests growing on the newly abandoned farmland. (For scale, the world's forests are currently estimated to hold 296 gigatons of carbon in total.)

After the ravages of the Black Death finally died away, on the other side of the world another tsunami of disease would be unleashed toward the end of the fifteenth century, the worst of them all. In October 1492, Christopher Columbus made landfall in the West Indies. In doing so, he opened a floodgate, releasing a rushing torrent of biological interchange between Old and New Worlds in what has been termed "the Columbian Exchange."

The mixing of heretofore separate biotas began globalization in earnest. Tobacco, tomatoes, and potatoes were among the crops brought to Europe; sugar cane, coffee, and wheat were introduced to the Americas. And not all introductions were purposeful. Some, like black rats and earthworms (which hitched rides in "dry ballast") were stowaways on ships. Yet others, pathogens, were stowaways in the bodies of the human travelers. Diseases common to the Old World ripped through the Native American populations with terrifying speed, one after another: smallpox, measles, chicken pox, bubonic plague, malaria, and more . . . (In return, transatlantic sailors brought syphilis home to their families.)

If the Black Death had been horrifying in its scale, this was another thing altogether. Nearly 90 percent of the pre-Columbian population of the Americas is thought to have been wiped out between Columbus touching down in 1492 and 1650—when only 6 million people are thought to have been left alive. (For a sense of how empty the continent must have felt, the present-day population of that same land mass is over a billion.)

In some places, the mortality rate was even higher. Hispaniola's 1492 population stood at around a million; fifty years later only a few hundred traumatized individuals were still alive. It is "probably the greatest demographic disaster ever," according to the eminent geographer William Denevan. Historians call it "the Great Dying."

Until then, the Americas had supported many advanced pre-

Columbian civilizations. In Amazonia, for example, the early European explorers reported in 1542 that in places the river was a busy thoroughfare, hemmed in on both banks by dense settlements of various tribal nations. The chaplain Gaspar de Carvajal recorded one town which "stretched for five leagues [18 miles] without there intervening any space from house to house" (for scale—the island of Manhattan is 13.4 miles long). The metropolis was ruled by a great overlord called Machiparo. "It was a marvelous thing to behold," wrote Carvajal. There were roads "like royal highways," fortifications, and at one point they visited a hilltop villa with porcelain tableware and candelabras "all glazed and embellished with all colors and so bright that they astonish."

When later exploration turned up neither hide nor hair of these lost cities, with their "glimmering white" avenues and plazas and canals, such reports were dismissed as works of fantasy. Despite great heroics among those questing for El Dorado, or the Lost City of Z—or whatever you want to call them—they were never seen again. Instead the rainforest stretched on, dense and dark, into which parties of explorers were swallowed whole.

However, recent discoveries have vindicated Carvajal many centuries after the fact: aerial surveys using laser sensing to study land shrouded by forest have revealed extensive networks of roads, houses, and pyramidal structures built by hitherto unknown cultures deep in the Amazon rainforest, in regions previously thought untouched by man. Fortified villages, garden cities, earthworks on a monumental scale, and—significantly—an artificial soil called "Amazonian dark earth" enriched with charcoal, manure, and compost, which would have allowed these peoples to farm in the thin, poor soils of the tropics.

Elsewhere, in the southwestern United States and Mexico, indigenous societies built dams and irrigation canals, and cultivated large quantities of maize, cacao, and fruit; while in what is now Florida, early Spanish explorers noted "great fields of corn, beans, and squash and other vegetables which . . . were spread out as far as the eye could see"; the Incas terraced the Andean highlands, and developed elaborate drainage and irrigation systems. Though these efforts were much less intensive than modern agriculture, researchers estimate that the collapse of farming and fire management in the aftermath of post-Columbian pandemics caused at least 193,000 square miles to revert to forest, woody savannah, or grassland, resulting in the uptake of an estimated five to forty gigatons of carbon in the space of a century.

Calculations of carbon sequestration—as with most questions of climatological modeling, and indeed almost all events of distant history—tend to be contentious. But, around the same periods discussed above, we do know that the climate was shifting. The start date is debated, but from around the fifteenth century, possibly as early as the thirteenth century, and well into the nineteenth century, Europe and America experienced a period of significant cooling known as "the Little Ice Age."

Some link the arrival of the Little Ice Age with changes in solar activity, and with volcanic eruptions (which we will touch on in a later chapter). But Ruddiman points to carbon sequestration as an—and perhaps the most—important factor. Perhaps all three played their part; either way, it does seem likely that the large-scale regrowth of forest had a significant impact on the atmosphere. Atmospheric carbon dioxide levels reached a nadir around 1610, a little over a century after Columbus first opened the portal between the Old and New Worlds, the timing of which aligns with the time frame of succession in abandoned fields, which take fifty to a hundred years to reach peak carbon storage.

When the early European settlers walked out into what seemed a vast and pristine wilderness, thinly peopled, it seemed to them to be an empty place, God-willed, ready to be claimed. They cast themselves in heroic roles, struggling against a wild and untameable nature. And even today, in popular culture, a vision of the primordial wilderness—the unsullied Eden—that preceded modern America persists: a place where deer sprang nimbly through the tall grass and darted through endless forest. But what has grown increasingly clear is that much of this "forest primeval" must have been relatively recently regenerated. The Arcadian dreamscape celebrated by the colonial pioneers was, in fact, a post-apocalyptic one.

Here in the former Soviet Union, the proximate cause of the abandonment is very different: politics, not pandemic. The net result, however—the pattern of desertion, regrowth, and carbon sinking—is not so dissimilar. And indeed, in terms of the square footage of abandoned farmland we are dealing with, worldwide and simultaneously, this is on a scale never before seen.

Estonia, early evening. We drive inland in the gathering dark, through wide flats and unfenced fields showing the unmistakable shaggy countenance of mid-stage succession. Young trees huddle together in cliques; shrubs and bushes congregate along streams and road shoulders.

We pass a series of huge, long warehouses of white brick, with steeply pitched roofs. They seem absurd in the rural setting, like spaceships. "You want to look?" says Tarmo. We pull over.

A track fringed with faded weeds leads us to the front of the nearest behemoth. They are enormous dairy sheds, emptied of cattle and most of their fittings, but a few old troughs and partitions remain. The ceiling is low and beamed, supported by two rows of concrete columns, which lends the interior the feel of an open-plan office. Old stalls, where the cows once stood ready to be milked, wear a thin chiffon of lichen or algae. The wide wooden sliding doors have come away from their hinges and hang at odd angles.

In a back room I find an old straw mattress on the bare floor, ripped open and spilling its stuffing. About a dozen matchboxes and cigarette packets lie crumpled in the corner among some broken glass, some lengths of piping, and an empty box of what looks like prescription medicine. It's clear someone has been sheltering here, not so long ago. I step back, shut the door.

The yard between this huge brick hall and the next has grown shoulder-height with flowering weeds, now gone to seed, and a few slender birches letting gold leaf flutter to the ground. Inside, the air stands still and cool.

Here's a concept I stumbled on once: "the ghost of herbivory past." As with so much in science, it has a meaning at once prosaic and poetic. Browsing animals leave traces of their presence long after they have left: manure-enriched soil and plants uneaten over time come to dominate. A legacy of what has come before. It is a form, they say, of ecological memory.

We turn back to the car, and as we go, I feel those ghosts moving through the barn. Warm breath rises in clouds over hot flanks.

And outside, when the landscape is wheeling by my window, I see ghosts of old pastures too, and the forests of the future. Marching over the flood plain, down to the sea.

NUCLEAR WINTER

Chernobyl, Ukraine

APRIL 26, 1986. 1:23 A.M.

T he heavens are black velvet, shot through with stars. Liubov Kovalevska is in her bedroom in Pripyat when two flashes light up the sky from the east. Lightning, perhaps. Or shooting stars. There is a rumble like thunder, or fighter jets overhead. Not loud enough to really wake her. By morning, she will have forgotten it ever happened.

But in the sky: a glow. Soon, for those who are watching, a pillar of fire can be seen rising from the wreckage of the plant, the chimney glowing red hot and the walls split open; the air above and around it shimmering with heat and steeped in vivid color: red, orange, sky blue. The color of blood, one worker remarks. No, he corrects himself, like a rainbow. It is stranger and more beautiful than he could have imagined.

Later, in daylight, a crowd of residents gather on the railway bridge to watch the reactor burn. It is a warm spring day, unusually so: a heat wave. The blossom hangs heavy on the bough, leaves still unfurling in the trees under a cloudless sky. They are out in bare shoulders. But all the time, radiation is snowing invisibly down upon them from a great height. Falling, falling, falling on their bare, oblivious heads.

By early afternoon, armored vehicles and men in masks roam the streets. That something has gone wrong—badly wrong—is becoming clear. What, exactly, no one yet seems to know. Or rather, no one is yet willing to tell. Vehicles, not unlike fire engines, begin to spray the buildings with a soap solution, and then the road. White froth gathers in foamy mounds in the gutters. Children throw globs of it at one another like snowballs.

The next day, the residents of Pripyat city are informed by radio announcement that they are to leave their homes. Gather your vital paperwork and a change of clothes, they are told, and assemble in the street, where you will be evacuated by bus.

They leave, expecting to return in three days' time.

The apartment is empty, and the air inside stands cold and still. It has a damp, stale smell, with an undernote of something organic, of soil. Through the window, the low winter sun casts shafts of light through freezing fog.

In what was once a bedroom, now emptied of furniture, the floorboards are in disarray. The far end is upturned and tumultuous, a wave broken on the beach and retreating. Those boards that remain in place are pliant, spongy beneath my weight. On the opposite wall, mint-green paper is in the process of sliding sensuously to the floor, slipping from the plaster like silk and pooling at its ankles. The window has been latched but the glass is missing. A semicircle on the floor where the rain and snow comes through is damp and greening over with moss.

I leave the apartment and climb the stairwell, poking my head through open doors. Birds' nests balance in unlikely places: in fuse boxes, on bookshelves, in desk drawers. Sprays of ferns sprout in damp corners. Paint demonstrates a thousand different ways of peeling, flaking, curling, crumpling to dust. Masonry and glass crunch underfoot. The linoleum of the stairs is coming away in sections from the tread, wet and flimsy, like the skin of a rotten apple, the skin of a corpse. The handrail wobbles in my hand. I make it up three flights of stairs before I lose my nerve.

The ease of access has surprised me: at some point between the evacuation and my arrival every door in the city seems to have been unlocked, swung wide. Trapdoors lie flung open, literal death traps, though some of them have been plugged by thin green branches stretching through toward the light. Lianas, twisted and taut, rope up external walls and force their way through broken windows. In theory, visitors to the city are banned from entering the decrepit buildings—soldiers cruise the streets in SUVs, watching for rule breakers—but in practice once you've disappeared through a doorway into a labyrinth of empty rooms you are impossible to catch and I'm not sure they're really trying.

In the middle school on Sportivnaya Street, the parquet floor-ing of the assembly hall lies shattered into a thousand pieces; a classroom lies ankle-deep in scattered textbooks, its desks a tan-gled barricade along the far wall; in an upper hallway waterlogged plaster flops from walls that buckle inward, where albino stalac-tites six inches long hang dripping under the joins between con-crete panels. Intact windows are misted with limescale. Where the plasterwork has come away, a loose-weave hessian layer of roots is revealed underneath. The new forest that invades the city rumples the road, pushing its roots under the asphalt like limbs beneath a bedspread. ·

Seventy percent of the zone is now forest. Pripyat is the ter-ritory of the birch and the maple and the poplar, and their leaves lie in a thick litter on the tarmac, branches bare and colorless but for the globes of green mistletoe and the mustard lichen that mists their bark. Matted shrubs crowd their lower reaches, flecked with the red points of rosehips gone soft. Ivy weaves between their legs. Apartment blocks rise like concrete islands from a sea of green. The trees are standing too close, crowding doors and blocking windows, growing in tight against the walls. It makes the buildings uncomfortable to look at: one feels keenly the infringement of their personal space. Saplings lean precipitously out from upper balconies. Creepers climb signposts, balancing awkwardly on the upper edges with nowhere else to go. They seem desperate, as if scrambling to escape from rising floodwaters.

A high, elephantine scream sounds from somewhere in the woods not so far away. A heavy iron gate, turning on old hinges. Air blown hard through metal pipes. After a moment, I place it: I've heard it before, in a very different context. Elk. They will not harm me. But still, the cry makes me nervous. They are too close. We are in their territory. I feel like a lost child wandering in the wilderness, far from safety. Even here, in the center of town.

Two months before the meltdown at Chernobyl, the minister for power of Soviet Ukraine had assured the press that the nuclear plants there were very, very safe. They'd run the numbers, he said: the odds of a meltdown were one every ten thousand years. This, everyone agreed, seemed like a good trade-off. A relatively small risk for a lifetime of clean, efficient energy.

On the other side of the Iron Curtain, the Americans were making similar calculations. They were a little more cautious: according to their latest estimate, the odds of severe core damage at a plant was three in ten thousand. Still low perhaps—but in reality, as the partial meltdown at Three Mile Island in Pennsylvania a few years previously had shown, nuclear incidents could and would happen. In fact, if you ran the sums for the *accumulated* risk of multiple plants, over multiple years, risk rose considerably. For a hundred reactors over a period of twenty years, that figure would rise to a 45 percent chance of meltdown.

Worldwide, since the first reactors were built in 1954, there have now been two nuclear disasters rated as a seven—the highest grade—on the International Nuclear and Radiological Event Scale.

The 2011 disaster at Fukushima in Japan is the most recent, caused when an earthquake and subsequent tsunami caused cooling systems to fail, producing explosions and three partial meltdowns. The incident was serious enough that the evacuation of all Tokyo was considered at one point, and a 140-square-mile exclusion remains in place.

And, beyond nuclear reactors, there are many other locations that have been forcibly sequestered due to radioactive contamination. At Mayak in Russia's Urals, for example, an explosion at a nuclear waste facility in 1957 saw a plume of radioactive dust drift over an estimated 8,800 square miles. An area of 38 square miles near the release site remains fenced off as a "radiation reserve," and no public access is permitted. At Hanford in Washington State, a former plutonium production plant, a massive quantity (685,000 curies) of radioactive iodine was released into the local environment during a push to produce fuel for nuclear weapons during the Second World War. It, and the 586-square-mile reservation that surrounds it, remain cordoned off and are still home to more than 50 million gallons of high-level radioactive waste, several tanks of which are known to have been leaking for at least fifteen years.

But Chernobyl is the most contaminated site of all. Though the explosion at its fourth reactor had only a fraction of the force of the atomic bomb dropped on Hiroshima, the nuclear fallout it released is thought to have been four hundred times greater, thanks to the huge quantity of nuclear fuel housed within the damaged reactor. Two people were killed outright in the hours following the explosion. Twenty-eight more died in the first few days from radiation poisoning, of one hundred and thirty four who were hospitalized with radiation sickness. An estimated two hundred thousand of the so-called liquidators working to clear the area were exposed to at least five times the internationally accepted maximum annual dose of radiation.

In the nearby environment, all forms of life were affected—often in strange or gruesome ways. Pregnant animals miscarried, their embryos dissolving. Horses four miles from the plant were killed as their thyroids disintegrated. An entire forest of pine trees was scorched rust-red, dropping their needles and then dropping down dead. Freshwater lake worms switched from asexual to sexual reproduction.

After the initial evacuation of Pripyat town, the whole area was shuttered: 1,600 square miles, an area bigger than Rhode Island, encompassing two major towns and seventy-four villages. It has several names, this place. The official title translates literally as the "Zone of Alienation." Others also know it as the "Dead Zone." It is the most radioactive environment on Earth.

These radioactive hinterlands are the consequences of human folly, hubris, of deals made with the devil. That they have been badly contaminated is evident, but it has also been growing increasingly clear that the Dead Zone is not dead at all.

The village of Paryshiv lies a mile down a deeply rutted track that cuts through dense shrubland and open pasture grown thick with silvered grass half-flattened by the wind and the rain. Most of the wooden buildings are in an advanced state of collapse: they come apart at the seams, walls falling outward; others stand open to the elements, the thatch of their roofs rotten and their rafters bare. Overall, the effect is that of a hurricane just blown through.

Ivan Ivanovitch's cottage is somewhat sturdier: brick-built, white-washed, with a corrugated roof patched with snow. The front door has been painted a cheery turquoise and the window frames a sky blue, although they are flaking at the edges. Embroidered net curtains hang limply behind the windows. The yard out front is packed dirt. Wire has been strung overhead between tumbledown outhouses for use as clotheslines. Everything is in a state of disrepair—although it is not uninviting. Black hens peck between my feet.

Ivan is shorter than me, bent over with age, and wears a tatty quilted jacket, oversized black boots, and a hat with earflaps tied over the top of his head. He seems excited to see us and talks non-stop in Ukrainian, as I smile and try to nod in the right places as Ludmilla, my guide, interjects when she can with translations. "He's lonely," says Ludmilla. His wife died eighteen months ago. Until recently, they had five neighbors in the village. Now, there are only two left. So he likes to have visitors. Lucky for me; not all the zone's residents do.

Ivan's late wife, Maria, grew up in this village and spent almost her whole life in the area, he says. He came to the region to work as a guard in Chernobyl town, and joined one of the collective farms. After the accident at the nuclear plant, they, like everyone else, were resettled: the Ivanovitches were allocated temporary accommodation on the outskirts of Kiev. They were told they would be able to come back "soon."

So they waited. It was a grim time. They worried about the live-stock they had left behind in Paryshiv, struggled to trace close family and old friends. It was too different from the life they had been used to. So, in 1987, only a year after the disaster and with the newspapers still full of the dangers, they decided to come back.

At first it was difficult. They, like the thousand other *samosely* (literally, "self-settlers") who returned home illegally, had to walk home and thereafter keep under the radar of the soldiers who patrolled the exclusion zone. But, says Ivan, he knew the guards, so they didn't give them too much trouble. Strictly speaking, they were not meant to be there, but in 2012 the Ukrainian government announced that they would turn a blind eye to the presence of the—almost entirely elderly—unofficial population of the exclusion zone, even going so far as to send regular health visitors and to deliver their pensions. There is electricity still, though no running water. Ivan even has a television, he tells us proudly, although admittedly it's been broken for a while now.

They were lucky in Paryshiv. Thanks to the leopard-spot patterning of radiation, their land was not *so* contaminated. The Ivanovitches kept cattle and chickens. They grew vegetables—although wild boar would come and dig them up.

What other wildlife? I ask. He laughs. Lifts his hands in an expansive gesture that requires no translation: try me. Wolves? He says: Sure. Too many. Gestures to the high fortress-like fence around the property he had to build to keep them out. He hears them, hoarse-throated, howling, through the long dark hours of the lonely nights. A few months ago, he says, the field near the house was frequented by a female wolf raising a litter of pups—he saw them running every morning in the dawn light.

The reality is, wildlife abounds in the zone. No one knows exactly how many deaths took place in the immediate aftermath of the accident, although it's assumed that in the worst-affected areas (like the Red Forest, those rows of scorched, broken trees) that radiation levels were enough to kill every mammal present within a few hours or days. But after a few seasons, the regrowth began in earnest. Animals reappeared: lynx, boar, deer, elk, beavers, eagle owls—and on, and on. Many of the species were rare, their numbers dropping all over the rest of the Soviet Union, but found sanctuary in the forest and abandoned farmland that makes up most of the exclusion zone. A decade later, every animal population in the zone had at least doubled in number. By 2010, the wolves had increased sevenfold. In 2014, brown bears were spotted in Chernobyl for the first time in a century.

The apparently remarkable recovery of the environment around the reactor—despite its injurious levels of contamination, even today—could be considered a provocative thought experiment made real. As James Lovelock has suggested, perhaps this "unscheduled appearance" of wildlife may indicate that aversion methods—the more frightening and insidious the better—could be the most effective method of keeping people out of nature reserves.

Other, similar, cases might include the Vieques National Wildlife Refuge on Puerto Rico—a tropical wonderland of eighteen thousand acres, whose bioluminescent waters harbor rare species such as the green sea turtle and the West Indian manatee, while its lush forests shelter one hundred and ninety species of birds and nine species of bat—which simultaneously doubles as a quarantine for the enormous quantities of unexploded ordnance, napalm, depleted uranium, and biological weaponry dropped upon the site when it was the U.S. Navy's Camp Garcia. Large tracts remain classified as a Superfund site and thus off limits to visitors.

In Colorado, the Rocky Mountain Arsenal, a former chemical warfare facility producing everything from sarin to mustard gas, was earmarked as a wildlife refuge after bald eagles, then an endangered species, were seen nesting there in 1986. Prairie dogs, mule deer, hawks, owls, and coyotes were also making their homes in the abandoned fields that had been cordoned off since the Second World War, and thus protected from the steadily encroaching urban sprawl of Denver, which now surrounds the site on three sides. Bison were later reintroduced, and the wildlife seems to get by despite the dangerous chemicals still believed to be leaching into the groundwater. The same goes for the buffer zone around the old Hanford plant; the stark shrub steppe skirting the contaminated site is also home to bald eagles, plus great blue herons, white pelicans, and porcupines, and was declared a U.S. national monument in 2000.

But Lovelock went further still: we might *choose* to contaminate land, thus creating exclusion zones voluntarily, as a perverse method of protection. Perhaps, he mused, "small volumes of nuclear waste from power production should be stored in tropical forests and other habitats in need of a reliable guardian against their destruction by greedy developers."

So is Chernobyl radioactive wasteland or safe haven? The answer is: both. In the immediate aftermath of the nuclear accident, there was a huge spike in ionizing radiation. But many of the radioactive elements released were highly unstable. They self-destruct, sometimes in seconds. Others over weeks. The most feared product of nuclear fission, in terms of its health impact, is iodine-131, which is easily absorbed by the body. Iodine-131 is stored in the thyroid, where it emits harmful beta radiation, damaging flesh in the immediate area and causing either the gland's destruction or, in lower doses, cancer. (At least four thousand cases of thyroid cancers among the children of Belarus, Ukraine, and Russia have been attributed to the radionuclide's effect.)

But iodine-131's half-life of only eight days means that its radioactivity decayed to a sixteenth of its original level within a month or so—and it has kept falling at the same rate ever since. By the mid-1990s, the total radiation level in the zone was more than a hundred times lower than in the immediate aftermath of the accident. In most of the zone, radiation has now declined to levels similar to what one might experience over an airplane journey, or during a medical scan. Today, most concern centers around the radionuclides cesium-137 and strontium-90, both of which have a half-life of around thirty years, and are readily taken up by plants, thus making their way up through the food chain. As a result, the flora and fauna of the zone have themselves become radioactive—more than half their exposure to radiation being internal, that is, emitted from their own bodies. Much of the produce of the zone (mushrooms, berries, fish, boar meat) is considered too dangerous for human consumption. But those organisms are not necessarily too damaged to *live*.

Where radiation accumulates: lichen, pond scum, the shells of snails and mussels, birch sap, fungi, wood ash, human teeth.

It will take another 270 years before radiocaesium and radio-strontium decay to relatively safe levels. The effects of this long-term, low-level irradiation are far from clear, and have been the cause of something of a scientific dustup. One vocal scientific faction, led by Anders Pape Møller of Université Paris-Sud and Timothy Mousseau of the University of South Carolina, have raised the alarm, highlighting a plethora of frightening abnormalities found in the animal populations of the zone: higher rates of cataracts in songbirds, albinism and tumors among swallows. Butterflies, spiders, grasshoppers, and bees are missing from the most contaminated areas, they say, and fallen trees and leaves are decomposing at a worryingly slow rate. Cut down a tree and find the accident etched into its core: before, the tree rings are widely spaced and pale; afterward, the wood looks orange, its layers densely packed from poor growth. Overall, they claim, there are fewer animals in the most contaminated areas, and the animals that are there live shorter, more desperate lives.

Most other scientists take a more measured, even cautiously optimistic, view—sometimes in direct contradiction to the formers' findings. One team set up by James C. Beasley of the University of Georgia, for example, cataloged fourteen species of large mammals in the zone using remote cameras, and found their distributions were not suppressed even in extremely contaminated areas. Similar studies have found no reduction of wild boar or rodents, indicating a surprising resilience in the face of long-term radiation exposure. In other words, though the radiation does them no good, the benefits of the absence of humans in fact far outweigh the harm. There is even a school of thought—albeit a marginal one—that proposes that low levels of radiation could even be beneficial, making organisms more resistant to damage and disease by stimulating DNA repair or immune responses: the "hormesis hypothesis."

That wildlife has returned to the zone en masse seems inarguable. Within moments of passing through the first checkpoint, I see three roe deer (stock-still beneath a snow-tipped bower, breath condensing in clouds at each nostril) and feel the place to be burgeoning, dense with life. Ivan sees boar and wolves in the long grass, where he had never seen them before. But gauging the damage is difficult. What we don't see are the stillborn, the stunted, the mutants that have died and been eaten before they have ever been observed. And even when we can find some—the double-headed pines I come across at Yanov, their torsos twisted, and limbs braced against each other, conjoined twins at war with their siblings—those individuals don't tell us much.

Mutations occur naturally, as do cancers: the question is how *often* they occur. In the early years after the accident, health services braced for a deluge of leukemia cases and mystery conditions among the local human population. Everything and nothing was attributed to the effects of nuclear fallout—children born with deformities of all kinds were pictured in the press; charities were set up in their name, thanks to a huge public outpouring of guilt and regret. But according to the World Health Organization and the UN, an uptick in (mostly treatable) thyroid cancer cases and cataracts is the only health outcome which can be ascribed unequivocally to the radiation. The other cases, however distressing, are not yet attributable to the radiation with anything approaching certainty.

At the same time, it's also hard to say with certainty that they *aren't*. Many remain skeptical of the international organizations' optimistic conclusions, and of the data they are based on; there has never been a large-scale, long-term epidemiological study that has settled the question.

The psychological impact of nuclear contamination, however, has been great. A loose constellation of disorders known as "Chernobyl Syndrome"—suggested to be psychosomatic—is common among the affected population, and whatever its cause is a source of real illness and suffering. The 116,000 who were forcibly rehomed and a further 270,000 living in affected regions have been profoundly traumatized by the events of 1986. Many lost their homes and support networks, others lost sources of income when it became illegal to sell produce from the affected areas. Overall, poverty-linked "lifestyle diseases" and poor mental health pose a far greater threat to affected communities than radiation exposure.

Scaremongering, a lack of reliable information, and persistent local myth-making have also contributed to "paralyzing fatalism"— wherein residents of affected areas, believing themselves to be doomed to ill health whatever they do, make reckless life choices:

abusing alcohol or drugs, smoking heavily, and freely consuming the mushrooms, berries, and game from areas that they know still to be highly contaminated. What does it matter, goes the reasoning, if they have all been poisoned anyway?

I ask Ivan: Does he eat berries? Or mushrooms? Of course, he says. Both. He hunted too, and fished in the rivers. Now, though, his health does not allow him to do so. He is old. Stiff. He has pains. He is eighty-one.

Ludmilla and he confer in Ukrainian for several minutes. "I told him to go to the hospital," she says during a pause, "but he says he has no insurance, and he doesn't like the hospital in Kiev." There is some bad feeling, I am given to understand, because Ivan feels he should get a higher level of pension, but the authorities ask for paperwork he can't provide. They come to a conclusion. Ludmilla squeezes his arm. I see her slip him money as we leave.

It's getting dark. We reverse down the rutted track back toward the road. As we do so, another light goes on in the village.

The ground out the back of the Café Pripyat is sodden and shrouded by heavy, wet snow. This lakeside café was a hub of industry in the desperate days that followed the explosion at the reactor. With the fire still raging, and the threat of a second, far greater, explosion hanging over their heads, hundreds of liquidators were posted here to shovel sand, lead, and boron carbide into sacks to be dropped from helicopters onto the burning core.

Dark leaf litter is in the process of metamorphosing into compost, the top leaves black and shining like the carapace of a beetle. The building is concrete and of unusual brutalist design: V-shaped concrete pillars support a long viewing platform, the smooth outer surface of the cement coming away in waterlogged plates, baring the rusted metal girders inside. An organic process, ruination: these artificial structures are just as vulnerable to decay as we are—they need constant attention, maintenance, occupation. Our presence is their beating heart.

Degeneration is never far away. Their guts and bones and musculature hide behind only a thin layer of paint and plaster. A single damp winter, or fertile spring, and an unoccupied house will mildew and mold. Windows will cloud. The rigor mortis of the built environment will take hold: doors will jam, joints expand. It has begun. But the structure as a whole will hold out until the roof is breached, and the water pours in, and the rot begins in earnest...

I pass through a door, its handle broken off, into the main cafeteria space. The ceiling of thin plywood has long collapsed, and the planks that held it up pile messily upon one another, nails protruding. The board itself has gotten floppy and pliable as fabric, hanging in parabolic arcs, draped languidly over ribs of masonry and wood. The whole desolate scene is dusted with powder the color of peppermint cream, which softens the effect, smooths the ripped edges.

The whole room is dominated by an enormous stained-glass scene that takes up the entire far wall: a moon rising in the west, into a sky of electric blue and crimson; and in the east, a burning sun, haloed in purple and orange and gold. Around and between, four godlike women rise, in simple robes, cups over each breast: the seasons. Winter drops snowflakes from one hand and blows through a slender scarlet trumpet in the other; Spring sheds tiny, parcel-like seeds as she looks to the stars; Autumn holds a bough of yellow leaves, casting rain below; Summer's single, disembodied hand rests at her elbow. The frieze has been painstakingly crafted by the pressing together of thousands of strips of colored glass—though many of the panes have now shattered, spilling their contents like gemstones across the floor. They crunch underfoot.

I step outside, descend a set of steps, and as I do, from nowhere the dosimeter in my pocket raises its voice from the occasional background click to the crackle of white noise, before overloading and sounding its warning alarm—a two-note siren of a timbre and pitch that seems intended to fluster. A jolt of adrenaline hits my system. "What's happening?" I blurt out, embarrassingly panicked but unable to contain myself. The silence, the inodorousness, the invisibility of the danger, preys on my mind. I am blindfolded in a room of poisoned daggers. "What's happening?"

Ludmilla waves my panic away. "Hot spot," she says, and my fear evaporates with the label, in the way a grave medical diagnosis might sometimes bring relief. She points to the bottom step; liquidators returning from the reactor area kicked the dirt from their boots and removed their clothes here before getting into minibuses that parked out front. (These minibuses, too contaminated to save, would later be buried in vast mass graves lined with concrete.) The dust they left behind has left a radioactive trace which persists into the present.

I hold my dosimeter over what seems to be the hot spot's epicenter—a concrete slab shrouded in leaf mold—and watch the numbers climbing. It stops at 15.82 microsieverts an hour—around a hundred times the background dose. Ludmilla thrusts her hand out over mine: the dosimeter in her grip is bigger, more professional-looking. Its reading is different, higher. "It likes cesium," she shrugs, and though I don't understand, I feel myself jealous. "Can I hold it?" I ask. I take the wailing device. When I step away, it falls silent. I relax.

The sky is white cloud, too bright to look at. The air is still. I follow the steps down to the café's private jetty, from which I look out across a smooth curve of ice that lids the lake's surface, one that does not break when I drop a rock onto it. What I think about is this: how the past is imprinted upon the present; in every place, but never more tangibly than here. I navigate it blindly in the present, with a dosimeter as a torch.

The zone is full of hot spots. Sometimes they have been marked by small yellow lollipop signposts, whose triangular faces wear the nuclear trefoil, the international symbol for ionizing radiation, in red. These signs are planted throughout the zone, marking places to avoid. They're not dangerous, exactly, but you shouldn't stand there long.

At the city limits we pass another on the side of the road. "Can we stop?" I say, and the driver shrugs. I get out and cross to the shoulder by the sign and keep stepping until I hear the warble of the dosimeter. All I see is a field of thick, tussocky grass poking through a thin layer of snow. Ungrazed, unmown. Pine seedlings are bristling on the edges of the clearing, amid a shimmer of birch—thin upper limbs of russet and wine over silvered, glimmering torsos. A thin white mist hangs about a foot over the ground, ethereal, waiting to be parted. It has started to snow. Heavy flakes fall in slow motion onto my hair, my crown.

I shut my eyes, feeling cold fingers touching my face, and visualize the radiation washing over me, bathing me in a current of which I am only abstractly aware. Whose existence is a matter of faith. It is one thing to understand the concept in a lecture, I think. Another altogether to comprehend your body at the mercy of it. It takes a certain, mystical, frame of mind; one I am not unfamiliar with. I surrender myself to it, to its greater power. I feel my boundaries blur, grow indistinct. Gamma rays pass through me on their way elsewhere.

After a minute or so, I step away, and hear the siren fall to a crackle. I feel nothing. I think: I am not afraid.

PART TWO

THOSE WHO REMAIN

5

THE BLIGHT

Detroit, Michigan, United States

The church is redbrick, solidly built and comforting, with an ornamental flair under the rake of the roof and in the arches over the windows. The exterior is neat, the guttering and roof tiles done out smartly in scarlet. But the sign that should show the times of service has come away from the wall. Leafed limbs of various autumnal hues have come to rest shyly against the brickwork, or prostrate themselves across the stone steps that lead up to heavy wooden doors.

That's all that gives it away. After a while, you don't need much. To the discriminating eye, you may as well spray-paint it across the facade: GONE TO SHIT, COME ON IN. Some places they do, I guess. Not in so many words.

Anyway, it's true. Round back, through a weed-choked courtyard, the door is wide open.

Is it possible to enter a church—however desecrated the space, however long since you last raised your voice in prayer—and not feel the cold firm presence of the sublime? Here it is: streaming in through the windows as clean, clear shafts of light; suffused in the gray and dappled stucco; in the talc of dove-gray plaster that dusts the burned and broken detritus on the floor.

The ceiling is huge and curved, banded like the interior of an enormous barrel. But at its edges, the upper reaches of the walls are bare brick and red, or charred charcoal where a fire has, in the past, taken hold before dying away.

I tack a bridal course down the aisle, passing between heavy pews that sit at odd angles. Some have collapsed, as if exhausted, on the tiles. To my right, the ragged red velvet of the confessional sags from its rod, tide-lined where filthy rainwater has soaked upward from the floor.

You can feel it in the air: the emotional trace of past epiphanies, crises of faith. Funerals and christenings, confirmations, the comings of age. Sheet music that served as the soundtrack to it all has fluttered to the ground and lies damply in clumps, like leaves. The grand piano beached on its side on the floor is missing its lid, soundboard, and all of its strings. Most of the white keys are gone too, and the black hang askew, their long wooden action bared, which gives them the spent look of burned matches.

The curved apse, high above, is a rich cream, rimmed in verdigris where the paint has flaked away. At its center, a dove surrounded by a huge, sun-like halo has been picked out in gold. A small, circular hole through the roof lets the sunlight in, which shines as a spotlight upon the floor by the altar. I step into its stream, let myself be anointed.

In the old Catholic school across the yard, I pass down dark corridors crisscrossed with shadow and light. Doors open onto classrooms on either side: on one, they are high-ceilinged and peaceful, empty and washed with a blur of light through vines, which shift outside on the breeze like the ocean. On the other, chaos. Ceiling tiles drift knee-high upon the floor, their warped struts swinging dangerously low, eye-level.

A blackboard in a downstairs room bears instructions for a final class, chalked in stylish cursive. A date: November, 1983. *Always set your margin before you type.* It's a desolate scene. But not an uncommon one, here, in Detroit.

Detroit is a city shrunk from its shell; too big for the people that live in it. Once America's fourth-largest city, it has been in terminal decline for seventy years, its population reduced by almost two-thirds.

What that means, in practice, is that to drive through the city is to spin through streets and sometimes whole neighborhoods in a state of what looks like decomposition. Tens of thousands of houses stand empty and falling apart, shingles melting from roofs like hot icing, brick-effect tiles sliding from alignment, sharp-edged gaps where rotten buildings have been pulled like teeth.

Where this decay begins and ends seems clearly delineated. Here, one neighborhood is clean and well-tended; aspirational. There, only few blocks away, one passes through a cumulus of steam rising from a grate in the road to emerge into what feels a nightmarish inversion. The air seems wetter, the sky darker, the buildings somehow haunted. Roofs sag, hollow-backed, and walls slump against one another. Plants force their way through cracks in the boards, press their leaves against the glass from the inside. Broken chairs and old strollers block pathways through overgrown lawns. It is not there, and then it is. Here, it has a name: the blight.

Blight: an almost poetic evocation of the literal decay manifesting in the derelict houses that line the streets in certain parts of the city. Houses secured against intruders wear boards over windows and doors, giving them a disturbing appearance, as if their eyes are covered. Some buildings, on otherwise respectable streets, stand wrapped in vines, shrouded. In all, more than eighty thousand properties—mainly houses—are thought to lie vacant in Detroit. Some are boarded up, some open to the elements, others semi-inhabited by squatters who camp inside without power or running water.

Yet more—nineteen thousand in five years—have been demolished, razed to the ground by the city, the only evidence of their existence the foundations that track the empty lots, floorplans from a forgotten blueprint. Now, whole streets stretch on, block after block empty. Places where people raised their families, where infants were born and took their first steps, where pensioners passed humid summer evenings on the porch—gone. All that is left are the empty lots, sometimes dozens of them in contiguous blocks. Acres of what has come to be called "urban prairie."

To pass through one is to travel along an unmarked road, often buckled or hatched with the craquelure of porcelain, through what seems like fields gone fallow, a flat plain thigh-high in golden grasses. The past is there as an underlay: ornamental trees and shrubs—ghosts of gardens past—stand in tight huddles, tiny oases in vast wastes. Fire hydrants pop up unexpectedly in remote parts; solitary streetlamps stand alone in silent vigil. Here and there, an occasional house stands alone and windswept—a tract home or sometimes a semi-detached, cut from its twin. The little house on an urban prairie.

In all, upward of twenty-four of Detroit's one hundred and thirty-nine square miles lie vacant—an area larger than Manhattan. Some say as many as forty.

The city is a rough-stitched patchwork of all kinds of cloth, some rough weave and some bejeweled. Here: skyscrapers, galleries, a busy park. Then you turn a corner and narrow town houses press tightly together, holding one another upright, porches sliding adrift. Broken cars on bricks and landlocked boats with holes in their prows. Soft fall light comes tinkling through the leaves, which lie gathered in a soft carpet whose color fades in and out according to the trees' various states of erubescence. Maples burn like flaming torches against velvet-skinned planes. Golden-leafed aspens stand shivering, lighting up the empty lots.

Drive past an empty house and then another—blackened and burned out—and see a flashlight moving inside it.

For this is not an empty city, although it is one deeply marked by abandonment. Blight, of the kind that they talk about in Detroit, is a phenomenon not wholly physical: it is a shorthand for a pattern that unfolds in abandoned places, one that drives their progression. Blight is broken windows, listing porches, fallen beams. It too is a distillation of the ways in which abandonment affects the psyche of the humans left behind, as an insidious force that pushes them from their homes—a psychological current that they must struggle against, else lose their grip.

In Detroit, in "blighted" neighborhoods, the advance of abandonment takes corporeal form in the decay of the buildings, and its encroachment brings a chill to the streets. But the process is not linear. Abandonment advances and retreats; as one building falls vacant, another may be revived. Where the infrastructure of the community has begun to erode—physically, socially—efforts by those who live there fend off its approach. Ultimate victory is not yet assured.

The full details of Detroit's meteoric rise and painful collapse are complex—the waves of desertion, the ripples of resurgence—but an overall shape can be sketched out quickly in bare figures. In 1900, a year after the city's first car manufacturing plant opened, the population rested at 285,000. By 1950, with the automotive industry in full flight, that figure had rocketed to 1.85 million. Then came the slide.

The Blight

As the wave of industry broke and drew back, major car manufacturers shifted from centralized plants into smaller, out-of-town factories, and then overseas, and with them the workers and the money. A single-industry city can deflate fast. From its 1950s peak, the Detroit population has fallen, fallen, fallen—from 1.85 million to 1.5 million, then to just over a million, and to 713,000. At the last census, in 2019, the population skimmed 670,000. What haven't left are the houses they lived in, the churches they worshipped in, the schools they educated their children in, the factories they worked in.

The hulking shell of the Packard Automotive Plant—all 3.5 million square feet of it—runs for a half-mile off East Grand Boulevard, a honeycombed mass rising to five stories in places, utterly gutted. I find myself inside it, on a wet October afternoon, in the company of two investigators from the Michigan Humane Society, on a search for stray dogs. We enter as everyone does: through a hole in the cinderblock walls, and find ourselves in a labyrinth of dimly lit halls and tunnels, a warren where water seeps across the floors in mirrored expanses.

Elise and Dave are all business, dressed in black and wearing bulletproof vests; I step warily behind them into a courtyard where seedlings of all shades find purchase on joists and sills and in the piles of fallen masonry, twisting between the steel rebars that jut from lumps of concrete. We pass loading bays filled like swimming pools with trash—broken crates and empty bottles and scraps of tinsel and plastic tubs and old blankets, just about anything you can think of—and into a dark gallery loud with the dripping of water. Dave's flashlight beam arcs through the mist, leaving a trace on my retinas as we step gingerly through a grit of broken glass, shattered into sugar granules.

Once the most advanced car factory in the world, the Packard Plant once employed forty thousand people, but was shuttered in 1958, and now stands as a ruined city within a city; concrete columns sheared and fallen like the monuments of antiquity. Outside, the lanes between the various annexes lie deserted and lifeless, like a parody of a streetscape of another era.

Eight miles to the west, the former American Motor Company headquarters lies in a similar state. Around half the size of the Packard Plant, it is still a massive edifice, a three-story complex fronted by elegant, yellow-brick offices and an art deco tower.

Entry is totally unpoliced. I climb the marble steps and pass through carved stone arches into executive offices. The wainscoting is coming loose, curling like paper at the corners; wood-paneled ceilings shed petal-sized flakes of varnish onto boards. Elevator shafts yawn dangerously at every turn.

A long corridor leads to the old assembly halls, which are utterly ravaged. They have the look of cheap sci-fi horror: flexible foil ducting spilling from the ceiling; orange tubing snaking across ceilings, miles of it; scraps of what looks like cotton candy—asbestos insulation—litter the floor. Everything seems shredded, ripped into tiny pieces and scattered on the ground.

The complex was owned for a time by a local scrap merchant, who claimed he was going to turn it into a home for children; the whole time, he was stripping it for parts. (He was later imprisoned for environmental offenses related to the asbestos.) Scrapping is maybe the most important element of Detroit's ecosystem of abandonment. When a building falls vacant, you usually have about twenty-four hours before the scrappers get in and start fileting it of anything of value.

Once, on an island in the archipelago where I live, I was walking along a remote single-track road when I stumbled upon the freshly crushed body of a rabbit. Its eyes were still bright, its fur soft and dry. Two ravens examining its corpse warned me off with irritable croaks before setting about the body: beaks like penknives, slicing flesh with neat little snips from the bone. Two hours later, the ravens were gone, the remains inherited by hooded crows. They fluttered up like butterflies to reveal clean bones, only the tiniest remnants of meat remaining, a skeletal form lying prone where earlier had been a still-warm creature. Next would come a closer shave—the carrion beetles, the fly larvae—before finally bacteria would fizz away even those pale traces.

This is what I think about, when I walk the corridors of the AMC headquarters. And this word: *domicology*, the study of the life cycle of buildings. To become a domicologist, one must recognize, first of all, that there is one. Wrote Webster:

> But all things have their end:
> Churches and Cities, which have diseases like to men,
> Must have like death that we have.

These buildings were scrapped in more organized style than most, but the pattern will unfold ten thousand times over in this city. Even as I depart the old building, I hear the voices of others—strangers—echoing down the halls. The fine grain go-over; the scavengers moving in, come to pick over the scraps of the scraps. They are motivated, organized, mobile, hungry. They move mainly by night, transporting their wares by supermarket carts. If you ever find yourself in an abandoned building in Detroit, they will always—always—have gotten there first.

Oftentimes homeless themselves, or close to it, the scrappers—like the carrion crows—depend upon building-death for their subsistence, and in this way accelerate a process of decay that in less populated areas might take months or even years. First go the furnaces and heaters and water tanks and the wiring and the plumbing. Later, the aluminum panels that waterproof gable ends. Soon, the buildings have slipped beyond repair, and their bodies returned—via a roundabout route—into the cycle of resources, for $0.45 a pound.

But even after the scrappers have taken their fill, the bones of these superstructures remain, juggernauts on the city skyline. These are just two of Detroit's white elephants, which lie falsely in wait for a fresh purpose that may never come, their ribs picked clean. Now crumbling and collapsing. Perhaps past saving.

In Detroit, one becomes attuned to the various flavors of abandonment, in the way one might come to recognize the different species of trees in a forest. What distinguishes an empty property from an occupied one: a lifeless look about a house's face, a certain stillness of the air inside. And then, what marks the empty property from the abandoned one: the drawn curtains, the tottering pile of mail upon the porch (the lower strata in slow disintegration), the thin overlay of grime upon the windows and doors and steps—all are features common to both kinds. But in the abandoned, there too is a terminal look—a sense of sagging, of rigid spars gone soft. Of rising damp, encroaching rot. The pallor of the undead.

It is instructive, I feel, to consider the question of abandonment in the context of a city. To zero in on what, exactly, marks a property as abandoned, even when simultaneously surrounded by— even occupied by—people. The city of Detroit has had to develop its own definition, for administrative reasons: to be classified as abandoned, it must be both *vacant* and to display what they describe as "outwards signs of blight."

Blight, as a synonym for urban decay, emerged from the work of the Chicago School of Sociology, whose influential approach in the early twentieth century leaned heavily upon ecological models. Cities, the thinking held, functioned like any collection of living organisms. They too had a life cycle, and would evolve over time in certain predictable ways. The researchers turned ecological terms to new effects: neighborhoods experienced *succession* (as incomers displaced existing residents), for example, and *invasion*s (as various cultural groups displaced one another).

A city experiencing sudden, negative demographic changes might be characterized in epidemiological or even pathological terms:

as if a kind of contagion, a social disease, was taking hold, leading to withering and degradation. "Blight," an agricultural term dating from the sixteenth century, when it served as a catchall for sudden and devastating crop death ("any baleful influence of atmospheric or invisible origin," as the *Oxford English Dictionary* has it, "that suddenly blasts, nips or destroys plants"), became the favored descriptor. It's a vivid image, one that grows increasingly lurid the longer you think of it: blight like a mildew, a fungus, a black pox spoiling green and glossy foliage; blight like necrosis tearing through the fields, rotting potatoes where they lie in the ground, turning them corky and ruptured, riddled with holes and dark growths, rendering the worst a putrid, liquid mess.

To talk of urban blight, therefore, is to talk of a socioeconomic malaise drifting through the streets like a miasma, slipping in through the windows or the gaps under the doors. Ripping through neighborhoods like influenza. In some places, like the plague.

And with this image implanted in one's mind, it's hard not to interpret the worst of Detroit in that way: seeing blight where the roofs collapse inward, in the charred and blackened beams and the puddling of the rain upon the floor. It makes one uneasy to step into those dark, diseased buildings, whose chimney stacks slide from the vertical, their broken panes unmended. The thought of blight adds a new layer of perceived peril—not only from the physical danger of the derelict buildings, or the imagined ne'er-do-wells lurking inside, but the *baleful influence* that might spring from them, as if we carry the blight home in the fibers of our clothes.

What is true is that research has shown "blight" to be more than just metaphor. Urban abandonment *is* contagious, in as much as the dereliction of one house on a street makes it more likely that its neighbors will become abandoned themselves. Only a small increase in vacancies precipitates a huge drop in house prices; beyond a certain point, it stops making economic sense for owners to sink money into maintaining a property that will not hold its value—and housing in the area begins to degrade. Vacant properties—and those within a radius of one hundred yards—are also more at risk of fire.

And, just as vacancy attracts scrappers who accelerate the decay, the decay itself attracts crime. This is not just true of Detroit. Studies of Philadelphia and Austin have found that crime rates spike on blocks with vacant buildings, specifically violent assaults. An empty house makes the perfect shelter for fugitives, or those taking drugs, or for prostitution, and myriad other crimes. Detroit, which has the highest vacancy rate in the United States, is also the country's most violent city, according to the FBI. Dead bodies are discovered in abandoned houses in Detroit at a rate of around one a month; corpses are found hidden in bins, or burned in arsons; victims are shot, strangled, or tortured.

In Detroit—where the number of vacant buildings doubled between 2000 and 2010; where tumbledown clapboard houses are grown over by the feather-leafed ailanthus, the "ghetto palm"; where foxes, pheasants, and opossums have set up home in the thigh-high grasses of the urban prairie; where falcons nest on the roofs of abandoned skyscrapers and beavers reclaim the river bank; where coyotes howl at night in the city's west side—there has been a rewilding in both senses of the word.

This word, *blight,* billows up in conversation wherever I go. Blight slips between us constantly: unseen, and yet all-pervading. It takes hold of the mind and grips it. I had never heard the word in this context before—its usage is an American invention—yet I too begin to see it: slipping as a specter between houses. Locals speak of blight the way one might speak of a malevolent spirit that stalks the halls at night. It felt at once inarguable and indecent; fitting metaphor, and yet, when applied to a living community, almost impossibly provocative. Though many voiced its name, I found myself unable to do so in company.

What came to mind, whenever I heard the word, was a conversation I had once, at home in Scotland, with a researcher of public health. I met him in a hotel bar—a polished, empty space, quiet but for the clinking of glasses—to interview him on his work on urban decay and its insidious effects. I never wrote the article, but I still think about it—what he had found, what he had to say. The fact was this: they die younger in Glasgow than they do in other cities. They die younger in Glasgow than in Liverpool or Manchester or Belfast—all deindustrialized British cities, with similar histories and demographics and patterns of deprivation. They die younger, in all social classes, and younger than expected even when adjusting for unhealthy behaviors. They die younger in

Glasgow, and no one really knows why. This "excess mortality" as the researcher defined it, was more generally known as "the Glasgow Effect."

"It is as if," one writer ventured in 2012, "a malign vapor rises from the [River] Clyde at night and settles in the lungs of sleeping Glaswegians." A *baleful influence*, in other words, of *atmospheric or invisible origin*. But if blight really should be a disease—a disease at the level of the city—then perhaps there might be a cure.

In 2014, a presidential task force chaired by three prominent local leaders in Detroit declared: "Just like removing only part of a malignant cancerous tumor is no real solution, removing only part or incremental amounts of blight from neighborhoods and the city as a whole is also no real solution. Because, like cancer, unless you remove the entire tumor, blight grows back." They published a detailed inventory of Detroit properties, its neighborhoods analyzed "block by block, parcel by parcel," alongside a call to demolish forty thousand derelict or dilapidated buildings. The instructions were clear and practical: identify the blight and eliminate the sources of infection.

In this way, the city might hope to tear out the diseased parts of itself, as a gardener might prune a tree or a shrub, and hope that by cutting it hard back for the winter it might prompt a flush of fresh, sweet growth come spring.

Targeted action was important, the task force underlined. Resources were limited, and the problem deeply rooted. They reiterated a point I had often heard voiced in the context of Detroit: streets and neighborhoods might pass a "tipping point" after which it might become too blighted, to spoiled to save. It could only be cut off, amputated, pulled out by the roots so as to halt the spread, and save the crop. Tough decisions. But the results would be worth it. They quoted Socrates: "We can easily forgive a child who is afraid of the dark; the real tragedy of life is when men are afraid of the light."

And in a sense, they were only extrapolating from, making official, folk wisdom that had percolated through the city for years. In the years preceding the city's bankruptcy in 2013, when the streets had gone unswept and streetlamps unlit, local initiatives had sprung up in the void. John George, a former insurance salesman, turned activist in 1988 when an abandoned house behind his own turned into a crack den. He organized a group of neighbors and together they boarded up the place. From such beginnings his organization, the Detroit Blight Busters, grew, and over three decades they have demolished over nine hundred derelict houses (in areas including the notorious Brightmoor neighborhood, where John was born) and boarded up or repainted hundreds more.

Tom Nardone, a charismatic local entrepreneur, became incensed by reports that the bankrupt administration would close city parks to save money, and decided to take matters into his own hands. "It didn't make sense," he tells me over tacos in Mexicantown. "All it meant was they would stop cutting the grass and picking up the trash. I passed a small park every day on my drive home, and I thought, shit, I could mow that myself. So I bought a lawn tractor."

He and a group of "middle-aged petrolheads," as he affectionately describes his volunteers, formed the Detroit Mower Gang, and have tended not only to parks, but sports fields and playgrounds in abandoned schools all over the city, and various vacant plots and derelict sites. It was impossible not to be swept along by his passion, his scrappy vigor, his get-up-and-go.

After lunch, Tom drove me to a concrete velodrome built in the 1960s. It had become overgrown, but a few months before they had cleared it again for use by the community. He led me around the track proudly, showed me pictures from social media of kids on bikes, racing and doing wheelies. It felt, I thought, normal. Like any sports field in any city. But that was the achievement. In a city of derelict lots and urban prairie, the neat green lawn is the universal signifier of order.

What Tom learned: mow any type of vegetation three times, and it will turn into grass. "It looks," he said proudly, "like we planted it that way." Succession in reverse. The Chicago School would find a message in that.

Constance M. King has lived almost her whole life in this pretty, primrose-yellow clapboard house in Detroit's North End. This was the house her parents took her home to after she was born—born, as she says, "in Detroit, Michigan, 1949."

It was a good neighborhood then. There were shops running all the way up Oakland Avenue, "good black-owned businesses." There was a drugstore, a grocer, a fish market, a poultry house, a shoe store. A Saks Fifth Avenue. Apex Bar, where John Lee Hooker played his first gig in 1943. And the Phelps Lounge, once host to James Brown, B. B. King, Etta James. Constance's street was an unbroken line of houses. Working families, raising their children, keeping their lawns manicured.

Except, she says, it went to pot. Families moved out, short-term renters came in. Houses fell into disrepair; sometimes they were destroyed by their own tenants. "A lot of the schools got torn down. There weren't many children left." There were a lot of drugs around, and those who took them. It stopped feeling safe to walk around the neighborhood.

Constance got married, and left the area. But the marriage didn't stick, and she soon came home to a neighborhood locked in steep decline. Her mother died, and then her brother was shot, just a few streets from the family home. Still, though, she couldn't bear to leave. "After my brother got killed"—she pauses—"I felt I had to start watching myself." She stopped shopping in the neighborhood, didn't walk to the shop or get out of her car. Not that there were so many businesses left to visit anymore.

But, this was home. For a time a local church was offering families a few thousand dollars for their homes; trade them in, and start again. But Constance found she couldn't bring herself to do it. "You see your family working—my mom working after my dad passed—to keep their property up, a roof over your head . . . you don't let it go down like that unless you really can't help yourself."

Plus, she knew her neighbors. They took care of her. Shoveled her drive in winter as she got older. Not that there were many neighbors, either. Steadily the street was emptying out. A neighbor on one side left in the 1990s. But Constance faithfully tended to the property, keeping the grass cut and the porch looking neat, "maybe some curtains or rags up in the window, where if somebody passed, they thought somebody lived there." Keeping it, in other words, merely vacant. Not abandoned.

Then one day someone got in there and realized it was empty. "I started hearing noises and: *bam bam bam*! They were tearing out the walls." Now it's just an empty lot.

Her house stands as one of four, which huddle together as if from rising waters. They are edged on either side by stretches of open grassland. A little way to the north, three matching houses stand shoulder to shoulder, the outer pair neat, their lawns filled with children's toys, their middle sister a warped and blackened shell with its lower windows boarded. At the end of the road, two wooden houses stand enveloped by trees, and peer between branches as if lost in a forest.

From Constance's perspective, blight seems less a disease than an all-powerful, indiscriminating force of destruction. A tsunami in slow motion. Or flood that must be fought, fended off. "How would I describe blight in my neighborhood?" she says, "I would describe the blight as unwanted. I would say, blight is just everything being destroyed."

Every day, she tends to her home, keeps the blight at bay. She tended next door too, until it washed away. But lately, she says, street sweepers started coming through again. They hadn't done that in years, not since she was a teenager. The North End is on the up, as she always knew it would be. "I'm just waiting for it to all come back."

The tide came in. In the end, she reasons, it must once more go out.

Later, before I leave Detroit, I climb back in the car and drive aimlessly through the city.

Streets scroll by the window: busy then quiet; high-rise buildings, then that rackety, rural air of the urban prairie. I cross the interstate, pass through a row of industrial warehouses, and find myself in Delray, west of the city center.

I feel a flare of recognition. Tom took me here earlier, to see a park he used to mow. But when we'd arrived it was gone. In fact, almost the whole neighborhood had gone—houses, parks, the lot, all cleared—ready for redevelopment as the foot of a new bridge to Canada and the plaza that will surround it. I trawl the roads that slice what is now an overgrown wasteland into squares, where tiny seedling birches raise their hands plaintively from between the bentgrass and wild rye. Haggard power lines sag and bow, some half-strangled by vines.

I turn down a side road, where the paving is cracked and crazed, greenery coming through like kintsugi. It's been abandoned longer here: full-grown trees have sprung up on either side, crowding the obsolete utility poles. The vegetation is thick and dense, crowding the track and coming close along the sides of the car. I see that, up ahead, a pleasure boat has been dumped or pulled across the road, blocking my way. I feel a shiver of unease and stop the car, reverse back the way I came, leave as fast as I can.

I drive on. I come to a row of buildings, rumpled and neglected, and halt next to an intersection. To my left, a huge square building stands alone, backing onto the river, which appears an eerie, artificial cyan. Beyond it rises the dark, dystopian shape of Zug Island, the realm of heavy industry—gas flare burning, tall chimneys belching fumes, coal heaps high as hillocks. The driveway is blocked, first by a bashed-up truck, polythene taped over its broken windows, and then by warped orange netting and a sheet with DO NOT ENTER—CRIME SCENE spray-painted across it in the color of blood.

The whole tableau—that mad-handed scrawl, the Mordorian backdrop, the glaring absence of police—sets my heart thudding. *Blight*, I think, feeling the word slide deliciously into place for the first time. I feel its presence pressing in on all sides. I want to leave.

I wonder this: Why would anyone stay here? Yet people do. ("It's nothing like it used to be," one resident told a local reporter. "But it is home. This is home.")

When researchers published the first paper on the Glasgow Effect, the unsolved mystery seized the public imagination. To the chagrin of the researchers, this unknown factor soon took on an agency of its own, a causative power to which premature deaths in the city—30 percent higher than might be expected—could be directly attributed. It was, as I told the researcher, a fascinating enigma.

I could see from his expression that he found my enthusiasm distasteful. Mystery achieved nothing for the city, he rebuked me. In any case, a new paper, soon to be published, now largely accounted for the deaths. One factor was proximity to abandoned and derelict land, which they linked to violence, contamination, and mental ill health. Another was more difficult to define.

It was this: Glasgow suffered greatly from invasive slum clearance and "urban renewal projects" during the twentieth century. Inspired by the theories of Le Corbusier and his utopian ideals, Glasgow bulldozed tenements in favor of tower blocks, and packed off the young and the fit to experimental satellite "new towns." All this, said the researchers, had ruptured the "social fabric" of the city: the matrix that supported the residents, held them steady. Morale fell. Health suffered. In came "diseases of despair."

The tearing out of "bad" areas in the name of progress had, in the end, torn the city apart.

In the United States too, anti-blight projects have a long and troubled history. Modernist architecture was rooted in the idea that if an environment was new and orderly, the community living in it would become so too. This needed a grand visionary at its head. ("The design of cities," as Le Corbusier wrote, "was too important to be left to the citizens.")

But so often grand visions come to fruition at the expense of the poorest, the most vulnerable. And, in Detroit, African Americans. It was overwhelmingly they who, in the 1950s and 1960s, were displaced from "blighted" homes in places like Black Bottom and Paradise Valley, and rehoused in high-rise, sharp-edged "projects." Whole neighborhoods were scrubbed away to make space for concrete estates, but though the buildings were new, the old problems remained. Redoubled, even. ("Poor creatures!" one critic wrote of Le Corbusier's future citizens. "What will they become in the midst of all this dreadful speed, this organization, this terrible uniformity? . . . here is enough to disgust one forever with 'standardization' and to make one long for disorder.") Henry Ford said that he wouldn't give a nickel for all the history in the world. But there is much that his generation taught us. If the goal of modernism was social revival, then it failed. In the tower blocks it became obvious that something deep and intangible and profound had been lost, something that had been there all along. It sounded like this: the rending of clothes, the tearing of cloth. Beware amputation. Beware the pulling out by the roots.

I think of Constance, smiling, hanging rags in the windows of her departed neighbors' homes. I think of John, boarding up the house behind his own. I think of Tom, tending to the grass in the parks. I pass community gardens, grown in empty lots. Murals of every color blooming across the sides of empty buildings. And in all these, the unspoken refrain: this is home, this is home, this is home.

If there is a cure for blight, this is it right here. Twice more, it will turn into grass.

6

DAYS OF ANARCHY

Paterson, New Jersey, United States

This story begins with a picnic. It was July 10, 1778. America's Revolutionary War was in full swing. General George Washington, the young Marquis de Lafayette, and Washington's then senior aide, Alexander Hamilton, were on their way back to camp after the Battle of Monmouth—a grueling non-victory that saw as many lives lost through heat exhaustion as from fighting—when they came across a most spectacular waterfall.

There, through a narrow chasm in the cliff face, steaming water plummets seventy-seven feet over the lip into a large basin below, "where it loses all rage and assumes the polish of a mirror," as an aide-de-camp wrote in his diary that night. From within the gorge they saw a fine spray rising "like a thin body of smoke" that drifted away on the breeze.

The general and his suite sat for an alfresco lunch under the spreading branches of a nearby oak, from where they admired the majesty of the falls, the rainbow colors that arced through the hanging mists, the thunderous crash of the waters. They dined on tongue, cold ham, and rum diluted with the waters of a cool spring that "bubbled out most charmingly" from the base of the oak. It was a most agreeable episode, they could all agree; a moment of relief after so many months of anxiety and struggle.

What struck Alexander Hamilton most forcibly, though, was not the clearness of the waters, where two dozen trout streams braided themselves into one, nor the breathy water meadows afroth with wildflowers, but the sheer unbridled power of the falls. When the war was won, in the years that followed, he, Washington, and their fellow founding fathers set about rebuilding the economy of a country now saddled with millions of dollars in debt. Previously strictly inhibited under colonial law, American manufacturing had never been allowed to develop, but now that his country had shaken itself loose, Hamilton found his mind returning to those roaring falls.

Where poets saw beauty and metaphor and natural wonder, Hamilton saw so much potential energy flowing away unused; a two-thousand-strong herd of horsepower gone unharnessed. Here was an awesome force with strength enough to drive mills and turbines.

In 1791, Hamilton formed what was essentially the first public–private partnership, and charged it with the transformation of the sylvan glades surrounding the Great Falls of the Passaic into something else entirely: a "national manufactory." Into the flesh of the hillside, channels and raceways and dams and sluices were carved so as to divert the rushing torrents from the falls' leap in the hungry mouths of the mills, and on their banks, the streets of America's first planned industrial city: Paterson, New Jersey.

It was, as the historian Richard Brookhiser has described it, the "Bethlehem of capitalism"; for Paterson, and all of her mills and looms and boiler houses and machine works, stood for something far greater than herself. She was a dream of a new future for the United States—one that came to pass, for better or for worse. It was the birthplace of American manufacturing. And later, its deathbed too.

I found Wheeler Antabanez online. He was an urban explorer of some renown, and I read his prose poems of the New Jersey underworld with a thrill of discovery. "I take comfort in decaying concrete and toxic waste," he wrote. "My goal is to find old factories, junked cars, discarded oil tankers and drowned boats . . . the underbelly of our highways, the dirty bottom of our sewage system . . . the forgotten zone." I figured we probably had some things in common.

Wheeler drives a glossy, all-black pickup, and wears black army boots, black combat trousers, and a black hoodie. He has a nervous habit of pulling this hood over his close-shaven head at intervals. We park up near the falls and pause to gawp as twenty thousand gallons of water tip into the abyss every second; a roaring, ferocious beast, as awesome as in 1778 but now constrained by a

Lilliputian clutter of pipes and concrete abutments. "The past above," as William Carlos Williams wrote, in his epic poem *Paterson*, "the future below / and the present pouring down."

Not a hundred yards away lies the former Negro League baseball stadium, now abandoned for two decades, its asphalt surface cracked through and fringed with silvered weeds, spindly trees coming up in rows upon the stands. We duck through a hole in the fence, walk out onto the diamond. A bottle-green scoreboard, its dark panels misted and defaced, overlooks the vestigial traces of the diamond, awaiting the next game. Another gap in the chain-link leads us to a steep and muddy track that rises to a rocky bluff looking out across the river and the industrial ruins at the heart of the city. Directly ahead of us, brick chimneys rise, elegant and columnar, thrusting incongruously from a forest canopy below, through which I glimpse at their feet the charred remains of the crumbling buildings.

In its heyday, the Passaic River powered three hundred and fifty Paterson mills, which employed twenty thousand workers. The city rode first booms and then busts: it was "the cotton town of the United States," then for a brief heady period grew into the locomotive capital of the world. Samuel Colt built the first repeating pistol here, and John Holland tested the first submarine. Industries arrived and departed; after each crash, the industrial district was born again. Its last incarnation, around the turn of the twentieth century, was as Silk City—its redbrick mills converted to hold looms and dye houses.

These overgrown seven acres at our feet represent a significant cache of industrial history—an extensive complex of old textile works, the former Colt gun mill, and the old and empty raceways that once fed them—but one now fenced off and hidden from view. Left largely to its own devices for decades, the former Allied Textiles site, as it's now known, serves as a potent symbol of the decline of American manufacturing, of the communities that once depended on it, and the toxic legacies of the industrial era more widely.

In 1945, the city corporation went bust for good. The waterways went dry, the factories became shuttered and overgrown. The tide went out, in other words, and left the population of Paterson high and dry. Now, almost a third of its hundred and fifty thousand residents live below the poverty line, unemployment is nearly twice the national average, and gang violence plagues the streets. Each year there are around twenty homicides; in April 2019, there were four shootings in the space of seven hours. Drugs are commonplace and easy to come by, particularly heroin.

Up on the bluff, Wheeler and I speak quietly so as not to disturb the encampment set up directly behind us: a single battered tent that might have been in place for months, and a camp chair—faded, looking out across the city. Over the ruins and across the past-its-best plaid of the Paterson streets and all the way to the gleaming skyscrapers of Manhattan, instantly recognizable and glistening with wealth.

We leave its resident in peace and follow a litter-strewn track down through the abandoned textile mills of Ryle Avenue, and across the river, looking for a way into the main complex. We pass the old Harmony Mill, now a Salvation Army depot, accessed via a concrete drawbridge over a dry waterway. Underneath, between the railings, someone has dragged a mattress that is bursting at its its seams, along with a watermarked pillow, and made an attempt to close off the entrance with netting.

In my search for abandoned places, I focused first upon the absence of people: this seemed to be prerequisite. But almost everywhere I look, I have found these "deserts" still peopled—by a skeleton cast of misfits or dropouts, peoples of the margins, and never more so than in Paterson, New Jersey. What I take from it is this: people might be abandoned too—abandoned to their fate, their own worst excesses. And it can work both ways: some choose to withdraw from society, permanently or for only a short time, for the freedom it affords them—abandonment in its other sense: a lack of restraint.

But with freedom comes danger. By stepping off the well-trod trail, whatever our reasons, we risk becoming lost. Worse, we might be unable to be found. Where better to consider the profits and the ravages of freedom than here: Paterson, ground zero of American capitalism.

Wheeler knows what he's doing. We duck through an opening that leads to a wooded track behind the old Essex Mill and into the unkempt estate where the crumbling ruins of four former mills and two dozen small outbuildings lurk unseen between the trees.

The path traces the edge of an old water channel, its floor hard-packed and dry, confettied with fallen leaves in lemon and pale green and tawny brown. About a hundred yards along we pass a small, shed-like building with a gaping front, its interior shrouded in darkness. Through the gloom I see mismatched crates and boards pulled together to divide the space. "Homeless camp," explains Wheeler, as we peer through the dimly lit space. There's no movement inside.

The ditch behind the shack is shin-deep in trash. There are empty drink bottles, food wrappers, aerosol cans, plastic beer cups, a running shoe with its laces missing, a child's stroller—and dozens of tiny resealable baggies of a sort that look like they once held heroin or crack cocaine. The detritus of lives eked out on scavengings and bodega bargains.

I step past a hypodermic needle discarded in the dirt, and a series of small pump houses, each deconstructed by different means: the first wrecked by a fallen tree, its cinderblocks lying scattered, as if across the nursery floor; its sibling burned out, roof rafters soot-black and rippling, split by the heat of the inferno.

We find the old Colt mill with its roof fallen in, its innards exposed, but the walls still standing and a brick chimney rising victoriously from the ruins, which are overflowing with trees and lush, leafy undergrowth. It's a strange kind of beautiful: the red-raw brick, the flush of leaves in autumnal array—climbing and cascading, draping luxuriantly. There are tiny, star-shaped wildflowers, trees bearing scarlet berries and amber ones too, that peek from between speckled leaves. Poison ivy unfurls coyly across the path, offering its treacherous hand. Light dapple dances over the ground.

Graffiti blooms on every surface, even the bark of the trees. It builds up in layers: older tags faded and painted over—bulbous letters in lime and mustard overlap a spiky hand design in pink and indigo, and above them stark copperplate picked out in white, the whole collage grown over with ivy.

I follow Wheeler into the dark heart of the place, into the mill itself—a huge turbine hall with the roof collapsed in, where enormous beams have swung inward to rest upon the shoulders of a row of metal girders. I have the feeling of standing inside the immense rib cage of an animal still breathing its last breaths; a monstrous coolant system, a mass of pipework, makes up its veins and nerves.

Just beyond, a wall perhaps two or three stories high has almost entirely collapsed, except for a thin upper arch of brickwork balanced unsteadily, supported from underneath by a spindly metal pipe. Leafy climbers rappel down through the gap, and tremble in the breeze like streamers.

As we advance through the complex, the paintwork upon the vast canvases of the factory walls grows increasingly hallucinatory. An elf-eared skull opens its mouth wide to reveal a doorway, through which a room serves as a garbage dump, waist-high in domestic detritus—sodden clothes, paint cans, plastic bags, grill trays. A blue-skinned woman pouts wine-stained lips and flashes the whites of her eyes. The glowing figure of a jade mermaid lifts one finger to direct us onward.

While William Carlos Williams was struggling through his *Paterson* sequence, he confided in one of his correspondents—the young Allen Ginsberg—that what he really wanted to do, was to "really go to work on the ground and dig up a Paterson that would be a true *Inferno*." In the bowels of the old mill, I wonder if I've found it. I step warily into a dark internal enclosure, gloomy and inch-deep in dirty water. Plastic vials of prescription painkillers lie open and empty in the pool.

But Ginsberg too hailed from this part of the world, "the same rusty county," as he put it in an earlier letter. He had learned to love his hometown. "Paterson is only a big sad poppa who needs compassion," he wrote in reply to Williams. "... I mean to say," he continued, warming to his theme, "Paterson is not a task like Milton going down to hell, it's a flower to the mind too."

We haven't seen anyone yet, but we aren't alone. I sense the presence of the inhabitants of this dingy purgatory, and occasionally hear the snap of a twig breaking underfoot not far away. More than once we come suddenly across makeshift shelters: a ruined outhouse, over which a tattered tarpaulin has been flung to form a crude tent, or a dusky pink quilt obscuring the entrance to a lair built inside a former store cupboard. A haphazard collection of clothes and belongings lie scattered on the ground outside: a white plastic garden chair; a cheap wheeled suitcase with the zipper busted; a leather armchair broken into sections. I avert my eyes, not wanting to pry, and we skirt the camp before passing through another doorway, before hastily withdrawing from the stench of urine.

We walk through a dark tunnel where a single shaft of light spotlights a tuft of yellowed grass. Close at hand, a figure appears suddenly through the gloom, striding toward us. A Hispanic man, the sides of his head shaved but the hair on top straggly and long, pulled back into a ponytail. Slate-blue tattoos etch both cheekbones and arms. He nods a silent greeting, before vanishing into the ruins.

"I think I know that guy," says Wheeler. "Okay," I say, doubtful, and we shadow the stranger up a vast rubble heap—planks, bricks, thick plastic sheeting, slabs, twisted metal girders, corrugated sheets, torn scraps of carpet—onto an upper deck, where light filters through pinholes in the rusted corrugated roof. The floor is flooded with stagnant water, milky with cement dust and reflective, shimmering with all the colors of the walls. The man is sitting on a step, rolling a joint. He doesn't seem surprised by our appearance.

"Cesar," he says, as introduction. His name.

He licks and seals the roll-up, offers it around. Up close, I can see his tattoos in more detail: on the side of his face nearest me, two musical notes etched level with his eye, and a stick-and-poke hatching along his hairline. He sees me looking and points out the rest: a lightning bolt on his left temple, the sign of the devil; a ship's wheel behind his ear; an anchor on his cheekbone. There's an upside-down cross on his knuckles, and a triangle of three dots by his wrist. "*Mi vida loca*," he explains. "My crazy life." "MS-13" has been spelled out across his bicep in gothic script.* He catches my expression and grins. "I'm not really a member. It's just . . . for fun." Right, I say. I don't know whether to believe him.

Cesar comes to the ruins to smoke. It's reassuring, he says, to have a place like this, where no one gives you any trouble. It supports a community, of sorts; a very loose one, in which its members arrive and depart without warning. Just being here is enough to indicate you have a certain amount in common. "I met my girlfriend here," he says, and shows us a photo on his phone: a pretty, addict-skinny blonde whose hair is streaked with color and the skin of her face embroidered in tattoos.

"She's amazing," he says. "But she's got . . . issues." He trails off. Recently she's relapsed into hard drugs. Drugs he won't go near. It makes her do bad things, act crazy. They're not talking. Sometimes, he says, couples drag each other down into the drug life. He gestures back toward the camps: those crumpled tarpaulins, the heaps of refuse. "Not me." When she gets sorted out, he'll be here, waiting.

After Cesar finishes his joint, he tags along. On this upper level, where roof panels have slipped from their holds and let in the elements, leaf litter has settled to form a mulch. Nettles and scrub oak and what looks like snowbrush are coming up in neat quadrangles, like a market garden. We slide back down the mound of

---

*MS-13 is a notorious criminal gang, whose members are predominantly from, or have roots in, El Salvador.

construction debris to where the rotten floorboards reveal the river as it rushes through the basement in a green-gray torrent.

Wheeler steps out onto a rusted metal girder over the water, leans out to inspect the rift in the building's foundations. I watch him silently. Hopeless to express caution in a place like this. Risk is inherent to the experience. To Wheeler, risk is part of the appeal.

Not far from where Wheeler grew up, there was an abandoned sanatorium where he and other disaffected teens could roam unfettered. They smashed windows, smoked cigarettes, "expressed themselves." It was a place to let off steam in a world that felt designed to constrain. Cesar feels the same way about this place, he says. He had a troubled childhood—born in El Salvador, and adopted by an American couple, he never felt he belonged. His brother found solace in the army and the routine it brought; Cesar went the other way. "I got a problem with authority."

This space—filthy and broken as it is—represents to both Wheeler and Cesar the kind of freedom they don't get anywhere else. George Monbiot once wrote of a "rewilding of the soul," and it is this same phenomenon that I sense amid the chaos and the grime. In a marginal space such as this, disordered and unclaimed, the unspoken expectations of society and its rules, however petty or important, fall away. To exist here is to renounce something. But it is to claim something too.

In an urban environment, entering an abandoned space is the nearest thing we have to stepping off the map. It offers anonymity, the succor of green space—without the order, the omnipresence of man, so implicit in the park or garden. An urban ruin might offer an effect upon the mind akin to slipping into the dark forest, or scaling a rugged peak, that same wild element, and we might seek it out for similar reasons. Amid the crumbling mills and towering, blackened chimneys—the skeletal remains of industrial giants—I feel a flickering, a stirring of the soul; the shadow of the sublime passes overhead.

Seen through this prism, the drug baggies, the bloodied needles, the spray-painted trees, even the grimy encampments, could be symbols not of a community in decline, but of an extreme anarchic liberty. (What Anaïs Nin said: "In chaos there is fertility.")

"There is more than one kind of freedom," as Aunt Lydia tells her captive audience in Margaret Atwood's *The Handmaid's Tale*: "freedom to and freedom from." In their new theocratic state of Gilead, the apprentice handmaidens are being given freedom *from*: freedom from objectification, freedom from certain kinds of danger. In the "days of anarchy," as Lydia calls them—meaning the before-times, *now*—"it was freedom to." I have never seen the latter so clearly epitomized as in the unkempt wilderness grown up around the old Colt mill. It's an instantly comprehensible demonstration, a defining of terms. There is no authority here to defy.

To be here is to experience a raw kind of exhilaration, laced with fear. There is a thrill in finding safe passage through a place with no safety rail; empowerment in the experience of self-determination in the extreme. But without the structure and strictures of normal society, the sense of *possibility* is so strong as to be dizzying. I could do anything, you think. I could be anyone. And no one will stop me. One realizes very quickly how thin the threads that hold our identities in place might be. Perhaps we only know who we truly are in a place where no one tells us what to do.

After an hour or two, we begin to walk, automatically, toward the way we'd come in. On a wall, upon every stone, a small work of graffiti has been picked out in a different shade: orange, sky blue, salmon, mustard yellow. The path—just a worn rat run through the shrubs—leads us past a derelict building, where the windows and floors have fallen in, and can be seen from above in a jagged heap in the basement.

Just here, says Cesar, only a few weeks ago, he had one of the most frightening experiences of his life. He'd been passing a summer's evening with a few acquaintances who also hung about the place. Most were heavy drug users, people he knew from around, people to have a few jokes with. One of them, a young man, bent to snort heroin, then collapsed forward onto his knees. The guy had overdosed—right there in front of him. "It was bad," he says now, looking pale.

The junkies he was with reacted immediately—by going through the man's pockets. Then they legged it into the bushes. "I couldn't just sit by and watch this guy die," Cesar continues. He doesn't hold with the cops, not normally, but . . . his voice trails off. He called

911, and an ambulance arrived. But it was too late. The man could not be saved.

Under a low branch near our feet lies a length of medical tubing and a crumpled white sterile package, ripped open, left over from the paramedics. "We had a sort of wake for him here after," he says after a moment. "And made, like, a sort of headstone . . ." He looks for it, but the memorial has gone. Maybe it has been knocked inward, onto the pile of debris, by someone who didn't know or didn't care what it stood for.

"Shit," I say. "Are you okay?" Cesar looks uncertain, and only hums a note in response. There is a quiet moment, and then he bids us goodbye and disappears into the brush.

Wheeler is quiet. I feel shaken by the story, by the dizzying sense of standing on a very high ledge and contemplating how far one might fall. This *freedom to* is liberating, yes, but with it comes the loss of protections, of responsibility figures, of safety nets, and that feels suddenly a hard trade.

On our way out, Wheeler and I cross paths with another resident of the ruins, a woman. She looks a little unkempt, unmade-up—though not unusually so. She wears a long coat belted at the waist and carries a battered handbag under one arm, apparently on her way home from work. I open my mouth to speak, but she glares at me so venomously, so obviously displeased by my presence in her domain, that I say nothing. Casting us one last scowl, she mounts the stairs, heading for one of the tarpaulin-covered shacks we passed earlier.

We follow the raceway back to the street. When we pass that first shed, with the gaping front, I see that a waist-high board has been drawn across the entrance, to block my view.

The hulking ruins of the mills along the banks of the Passaic may be the most obvious remains of the region's industrial past, but there are others more insidious. The poverty and the problems of addiction that I'd seen firsthand were two more. And in the water: another.

The dye shops and boiler houses and foundries and slaughter-houses of the city all produced waste, and all used the same method of disposing of it. The river that provided power to the mills also served as their waste pipe. At the beginning of the in-dustrial era, it seemed reasonable to assume that, once poured away, and then carried downstream, it might empty into the ocean and then be gone. The dye houses heated their solutions in vats, before dipping the skeins of silk to be stained. These vats emptied through pipes in the walls leading directly into the river; the day of the week, local lore holds, could be told by the color of the Passaic.

> Half the river red, half steaming purple
> from the factory vents, spewed out hot,
> swirling, bubbling.

The babbling trout streams known to the Lenape people, the region's indigenous inhabitants, soon became a stew of human effluent, industrial chemicals, and anything else that required disposal. A New York reporter traveling in the area in 1894 observed with horror, a "flood of filth" being poured into the Passaic—turning it into "a vile, inky fluid" which stained a dipped sheet of paper black. He followed "a terrible odor of putrid matter" to find the pool at the foot of the once-Great Falls filled with hundreds of dead perch which had "come over the falls and . . . died at once from the poisonous water."

By 1897, the sewerage commission reported that seventy million gallons of sewage was being dumped into the Passaic below the Great Falls of Paterson daily ("beyond its power to assimilate")—or around a third of the Passaic River's total flow. The fisheries had been completely destroyed. Manufacturing companies found the water so befouled that they could not even use it in their boilers, it added. Fish had disappeared from the river; a hundred riverfront properties in Harrison had been abandoned due to the stench.

By June 1918, the water was so choked with oleaginous industrial waste—oil, creosote, and acid, which bound together to form a thick film floating on the water's surface—that the river itself caught ablaze. *The New York Times* reported that the flames raced across the surface of the water "almost like an explosion, even singeing the hair and eyebrows of watchmen." The fire burned for hours, and a total conflagration of buildings along the banks was averted only when the tide turned and carried the whole flaming morass out to sea.

By William Carlos Williams's time it was, he declared, "the vilest swillhole in all of Christendom." Maybe it was. Unfortunately, there was worse still to come.

PART THREE

THE LONG
SHADOW

7

UNNATURAL SELECTION

Arthur Kill, Staten Island, United States

F arther south, where the waters of the Passaic finally meet the sea, there is a site that symbolizes, to me, the slow decay of industrial legacies. I pick up a rental car and head downstream, following Route 21 as it winds along the banks of the Passaic, toward Newark Bay. The road lifts and falls, on and off ramps rise and twist to meet it in midair; smooth, sculptural ribbons of road plaiting briefly together then peeling away.

The land here is densely developed, with that prosaic, stereotypically New Jersey esthetic of freeways and strip malls and vast flat parking lots. I pass from Paterson into Passaic, into the upper reaches of Newark—settlements marked as distinct upon my map, but appearing from above to bleed seamlessly into one another. *Conurbation*, I think, trying the word out for size. *Megalopolis.* I settle on *urban sprawl*—it captures the organic nature of the place, the way the apartment blocks and construction sites and depots spill across the land, how smaller buildings have sprung up in awkward poses that fill the cracks between the stadia and the gravel heaps and the warehouses.

It brings to mind the work of Robert Smithson, the sculptor and land artist behind the celebrated earthwork *Spiral Jetty*, whose work—with its focus upon decay, and ruination, and its perverse fixation upon "non-sites"—I admire. Smithson was born here in Passaic, and spent his early years across the river in Rutherford. And, like the sprawling, ill-defined settlements, I see his work as bleeding into or flowing from the Paterson poets. They brushed up against one another in life too: William Carlos Williams was Smithson's pediatrician; and like Williams—and Ginsberg too, in his own way—Smithson was preoccupied with the idea of finding a specifically *American* way of making art.

After a visit to Rome, Smithson returned to his home state and saw everywhere the warped reflections of the "eternal city" in New Jersey's "labyrinthine confusion"; its constant state of construction and deconstruction expressed an "almost Vorhazian sense of passage of time," as he would later remark. He wrote an influential *Artforum* essay, "The Monuments of Passaic," to this end in 1967, in which he described a tour of road bridges, pumping derricks, and parking lots and craterous building sites with all the breathless admiration of a tour guide.

The process of construction, he noted, looks a great deal like the process of ruination: "That zero panorama seemed to contain ruins in reverse, that is—all the new construction that would eventually be built. This is the opposite of the 'romantic ruin' because the buildings don't fall into ruin after they are built but rather rise as ruins before they are built."

It reminded him of Nabokov's observation that the future is nothing but "the obsolete in reverse"; the sprawl, the disorganization, the varying states of built and unbuilt-ness all contributed, he felt, to an architecture of "entropy"—that all-encompassing concept of disintegration and decay that so fascinated him, and would lead him to strip mines and chalk quarries and the rainforested former site of the collapsed Yucatán civilization.

Much of this later work sprung from the vista I look at now: a gap-toothed cityscape, the holes in the fabric of which constituted to Smithson monumental vacancies representing "memory-traces of an abandoned set of futures." It does feel a landscape of abandoned futures: where derelict mills and warehouses stand awkwardly along the waterfront, and clapboard houses in white and baby blue crowd between the ankles of bridges and overpasses, and stacks of disused shipping containers three hundred feet high tower over everything, and, away to the west, the sun going down.

Low tide. Early morning. I stand on the shoreline of the Arthur Kill, a tidal strait through which the waters of the Passaic—having passed through Newark Bay—drain away into the sea.

The ground feels swampy and insecure, strewn with beer cans and blue bottle caps and a scattering of narrow plastic pipes that must have fallen from a cargo ship. Salt hay and cordgrass grow densely together, their hollow stems dried to a barley-gold. Purpled seed heads shimmer in the breeze. Along a thin littoral margin the reeds, flattened by water, lie plastered to the ground like greasy hair. The foreshore beyond, a mud-brown sheen, gives off a thick stink of brine and decomposition.

It's worth it for the view. Ahead, only a few yards offshore, the wrecks of a hundred or more ships rise from the delicately shifting waters. The ruins are rust-red, spectral, and—lit with the roseate glow of the dawn—appear as an apparition: massed corpses emerging haggard from the depths, the ghosts of industry past.

If I squint, I can pick out their past forms and functions. The nearest, resting upright, was probably a tugboat: compact and chesty with an oversized smokestack. Beyond lies what appears to be a barge: low in the water, with the beams of its broad deck splayed loosely under broken rivets. After that, rank disorder. Car floats, steam ferries, navy tankers, fireboats—their working lives over, technology obsolete. I see portholed wheelhouses, twisted ladders, hulls rotting away to reveal corroded innards. Monumental in death.

The ships were gathered here over many years by the owner of a nearby scrapyard, and abandoned to the elements. By the time of his death in 1980, up to four hundred ships and boats had been amassed along the shoreline—an inadvertent museum of maritime design and a stark illustration of the "built-in obsolescence" of man-made objects Robert Smithson found so captivating. The scrapper's son, who inherited the business, recently made a start on deconstructing the graveyard, but it's clear that plenty of the ghost ships still remain. I gaze out to the ruins—the nearest beached upon the intertidal zone, their bellies black with estuarine filth, upper reaches the color of clotted blood. Farther out, an Atlantissian backdrop: the brine a smooth-blur mirror of the watercolor sky, breached by a thatch of spire-like masts and the church towers of tug chimneys.

Its existence—as one of many ship graveyards in the area—embodies the attitude toward waste disposal that was ubiquitous across the West through the through the nineteenth and twentieth centuries: let it sit. After a time, sometimes it becomes part of the landscape.

A few miles north, off the coast of New York's Staten Island, lies Shooter's Island. Once an oil refinery and shipyard—employing nine thousand workers during the First World War—it has been abandoned now for nearly a century. Aerial photos offer a time-lapse of its ruination. In the 1930s it is still sharply delineated, like a navy vessel sitting low in the water. By 1940, the boats that huddle along its shoreline appear exhausted and drunken, and the clapboard shingle shacks along the quaysides are slumped and squinting. By 1969, corpses of boats are piled disorderly on its western shore; jetties reduced to a stubble of wooden posts emerging from murky depths, the island's bare and muddy face wearing a thin sackcloth of rough grass. It was such an eyesore that a local politician suggested around that time that they should simply blast it out the water.

But in time, it grew in ecological interest. In 1980, egrets were discovered nesting there, and since then it has become a haven for waterfowl—glossy ibises, black-crowned night herons, and cormorants have all taken up residence in this shabby palace. Now owned by the Audubon Society, a conservation group, it is closed to the public to protect the birdlife from intrusion. The island lies hidden from view in the waters behind a self-storage facility, but satellite images show it bosky, densely vegetated, and untrodden with paths, its decaying boats seen faintly through the waters, stripped to their struts.

Farther east, in Coney Island Creek, another two dozen more wrecks sink into the mud. Cabins of ramshackle timber slowly up-end in the mire, their frames riddled with corrosion and caked with muck. Some are whaling boats, but most are barges scuttled after vast cargo ships laden with shipping containers rendered them obsolete. State authorities warn that to remove the wrecks would be to release the toxic chemicals buried in the sludge beneath them.

For it is this mud, or rather the poisons held within it, that is the true legacy of this region's industrial past. Robert Smithson asked whether Passaic—and by extension its namesake river, and the urban sprawl of its surrounds—had replaced Rome as the new "eternal city." And despite the apparent ephemerality suggested by the rotting warehouses and docks and bulkheads that line the lower Passaic and Newark Bay, the chemical memory traces of its past life may as well be eternal.

The oil refineries, tanneries, smelters, paint manufacturers, chemical and pharmaceutical works, and paper mills that proliferated in the area during the nineteenth and twentieth centuries produced noxious wastes of many flavors. Tanneries used sulfuric acid to strip hides, arsenic to preserve them, lead acetate to bleach them, and chromium to tan them. Hatmakers used mercury nitrates to turn fur into felt. And like the dye works in Paterson, they all dumped their waste straight into the water.

Later, factories produced polychlorinated biphenyls—better known as PCBs, the oily substances used as coolants or lubricants, or as electrical insulation, or hydraulic fluid, or in the manufacture of inks and adhesives and flame retardants—until their egregious impact on human health was realized in the 1970s. The insecticide DDT*—made infamous by Rachel Carson's iconic work of environmental literature, *Silent Spring,* and which kills creatures by attacking their nervous systems—was also produced at several sites on the lower Passaic.

Earlier, on my way south, I took a detour through Newark's shabby Ironbound district, where train cars rattle by on ground-level tracks, to see one of them—the former Diamond Alkali plant at 80–120 Lister Avenue, once described by Senator Cory Booker as "New Jersey's biggest crime scene." Originally a factory where cow bones were ground down to make fertilizer, the Lister Avenue plant was converted into a chemical works producing DDT in 1940, and later a manufacturer of phenoxy herbicides—specifically, the two chemicals that, when mixed into a 1:1 cocktail, constitute the notorious defoliant Agent Orange.

Despite the dangerous nature of its products, the plant seems to

---

*Full name: dichlorodiphenyltrichloroethane.

have been run with an appalling recklessness. At one point, the factory was pumping so much DDT-tainted waste water into the river that "mountains" of the insecticide rose from the shallows at low tide, and workers were sent out in waders to rake down the piles lest they attract attention. Spills would simply be hosed away with sulfuric acid (as much as thirty thousand gallons of it a day).

But the property's true notoriety centers around a by-product of phenoxy herbicide production: dioxin—an extremely toxic family of compounds, exposure to even tiny quantities of which, in any form, is carcinogenic. In humans, it causes every type of cancer there is, as well as stunting the development of unborn babies and afflicting wholesale damage to the human immune system. TCDD, the dioxin produced at the Diamond Alkali works, is the most toxic of all.

It was only years after production was halted that the public health risks associated with dioxins were fully appreciated. There is no truly "safe" level of dioxin contamination; it's one of the most toxic substances known to man. It is 170,000 times more deadly than cyanide. The U.S. Environmental Protection Agency considers water with dioxin levels of 31 parts per *quadrillion* (that is, 31 in 1,000,000,000,000,000) too contaminated to drink.

When, in 1983, tests revealed shocking levels of TCDD in the by-then-derelict Diamond Alkali works, it triggered a full-scale panic: the governor of New Jersey declared a state of emergency, closed the roads and tracks in the vicinity of the derelict plant,

---

*Full name: 2,3,7,8-Tetrachlorodibenzo-p-dioxin.

and bused in a dozen federal investigators in hazmat suits, to the horror of the local residents who'd lived there unprotected for years. Elevated levels of TCDD were found in the air-conditioning vents of neighboring buildings and in a local resident's vacuum; on the site itself, they found "massive" amounts: a reading of 51,000 parts per billion in the ground beneath an old storage tank.

The Diamond Alkali works were demolished and its remains entombed in a clay-sealed concrete grave along with 932 shipping containers full of dioxin-contaminated scrap removed from the site and surrounding properties. But though the Diamond Alkali plant has long stopped production, and factories like it are not permitted to discharge caustic chemicals into the water, a souvenir of the past remains in what scientists call "legacy contamination."

Unlike the crumbling artifacts admired by Smithson, there is no built-in obsolescence to these man-made chemicals. PCBs and dioxins and other "persistent organic pollutants" do not rot in the normal sense. In fact, they are virtually indestructible. Dioxin in particular: once accumulated in soil or sediment (or the bodies of living things), scientists estimate its half-life to be at least a century. Some, including the United States Department of Agriculture, go further, describing it as "virtually nonbiodegradable."

The production of PCBs was banned in the United States in 1979, in the UK in 1981, and in the rest of the world in 2001. Since the recognition of dioxin as a known carcinogen, production has been greatly reduced. But what has already been released will persist a long, long time.

This is the true legacy of these manufactories long closed, run by men long dead. Over the decades, these pollutants have settled like silt, putting down layers, marking off the passage of time, the evolution of human industry, by its bands of sedimentary poisons. Even if all humans were to die tomorrow, they would persist well into the future. A marker of our folly.

They are patient. They await dredging, disturbance, disruption. If stirred from their slumber, their reign of terror will continue. Hence the authorities' dilemma: lift the wrecks of the ghost ships, the industrial skeletons, that haunt the waterways of New Jersey and New York—and all those like them—and risk opening Pandora's box. Scrape up the sediment and incinerate it? Or leave it be?

On the banks of the Arthur Kill I wade into the intertidal zone, sinking up to my shins in post-industrial muck. The wrecks are beautiful, almost close enough to touch, but I know I mustn't swim in the water here. I've been advised to beware being splashed by it, even, lest a drop of it gets in my mouth.

Still, this water—toxic broth though it is—supports an ecosystem of a sort, albeit one much depleted. The waters of the lower Passaic and Newark Bay were once well known for their seed oyster reefs, though that environment was already destroyed by 1885 from all that "pouring of sludge, acid, and oily refuse into the waters" during the Industrial Revolution. Its vibrant array of fish life had largely died out or emigrated by the end of the nineteenth century—but a few pollution-tolerant species remain, and others are creeping back in now that the water quality is improving: bluefish, weakfish, blackfish, catfish, dogfish.

As with radiation in Chernobyl, those that live amid the post-industrial poisons carry the evidence of human impact built-in; over time, they accumulate PCBs and dioxin in their skin and their guts, in the fatty flesh along the line of their spines, until they have grown thousands of times more toxic than the water they swim in.

Benthic fauna, those bottom feeders that make their homes in the mud and the muck, are most exposed to the poison buried in the sediments: tolerant polychaete worms and soft clams and tunicate sea grapes predominate. And among them on the silty seabed scamper thousands of blue-claw crabs: olive-backed, about the size of your hand, and hiding underneath legs and bellies of a brilliant blue. Thousands of them. All the crabs you could eat. They look healthy enough. But a single Newark blue-clawed crab carries enough dioxin in its body to give a person cancer.

The New Jersey authorities hang signs along the seafront, in an attempt to discourage their consumption. The signs bear illustrations of the crabs, picked out in colors of cobalt and clay, struck decisively through with red lines.

DANGER!

DO NOT CATCH AND DO NOT EAT!

MAY CAUSE CANCER

There's a large, mixed immigrant population in this area, many of whom live in poverty and rely on the bay to put food on their table.

PELIGRO!

NO LOS PESQUE! NO LOS COMA!

CANCER

Many don't believe, or don't want to believe, or can't afford to believe the warnings. The crabs look so good, so glossy, so healthy. And their meat tastes so sweet.

危险！

禁止捕捉！禁止食用！

癌症

Dioxin is tasteless and odorless. And, local wisdom goes, if you locate and remove the green gland—the hepatopancreas—before cooking, you can try to avoid the worst of it. Still, it's a calculated risk.

위험!

잡지 마시오! 먹지 마시오!

암을

Get caught fishing for crabs in Newark Bay and you face a fine of up to $3,000. But if you're hard-pressed enough to be getting by through fishing in an industrial, rubbish-strewn estuary—this might be a risk that you're willing to take.

MAPANGANIB!

HUWAG HULIHIN! HUWAG KAININ!

KANSER

If they are so dangerous to humans, what of the fish themselves? The short answer is: it depends. The longer answer may offer us key insight into how the natural world responds to human impacts, and a glimmer of hope as to how life might adapt to survival in a despoiled, post-industrial world.

Before I continue, I should underline that there's no question that PCBs and dioxins have frightening, fatal impacts on life of almost all kinds. These pollutants have been observed to impair fish fertility, throw their hormones out of balance, and cause devastating deformities and developmental problems with their hearts, livers, and nervous system.

Through a process known as "biomagnification," those at the top of the food chain are the worst affected. In hot spots of pollution, such as the coasts of the UK and Brazil, the Strait of Gibraltar, and the northeast Pacific, killer whales are facing imminent population collapse due to PCB contamination. "Lulu," a well-known member of an orca pod resident to the waters off the west coast of Scotland, was found dead in 2016 after becoming tangled in fishing equipment. Tests revealed her to be one of the most PCB-contaminated creatures on the planet. She had also never ovulated; not a single new calf has been born to her pod for the past twenty-five years. Porpoises and dolphins are also known to be affected. In the Arctic, the bodies of Inuit people—those subsisting on traditional diets featuring seal and narwhal meat—have been found to contain such high concentrations of PCBs and other chemicals that they could be classified as hazardous waste.

PCBs have invariably negative impacts. Very few species can survive when they are present in an environment in any quantity. A small number of marine species, however, have shown an unusual resilience. One is the Atlantic killifish. The presence of this silvery, leopard-spotted fish, otherwise known as the mummichog or the mud minnow, in the polluted waters of the Passaic and in Newark Bay was first noted in the 1990s. Killifish are little slips of fish, which eat worms and mosquito larvae and serve in turn as foodstuff for a great many bigger fish. Despite their daintiness, they are extremely hardy. They don't mind if the water is fresh or salty, warm or cold, and in winter they simply burrow into the mud to escape the ice. NASA sent them to space, to see if they could swim in zero gravity. (They could. They spawned there too.)

And it wasn't only the Passaic. Killifish were popping up in other pollution hot spots. Because they don't migrate, the killifish are usually seen as an indicator species as to the environmental health of its home—a canary in the coal mine. Usually killifish are very sensitive to dioxin and PCBs, which interfere with embryo development. And yet, here they were. Although some were showing signs of physical distress—one study found that 35 percent of the killifish living in the creosote-soaked sediment at the bottom of Virginia's Elizabeth River had cancerous tumors—the fact that they were surviving in such places at all, never mind breeding, seemed remarkable.

In 2016, a paper published in *Science* pinpointed the manner by which the killifish had done it. A team of scientists led by the University of California, Davis, caught and genetically sequenced killifish from four contaminated harbors across the United States, including Newark Bay. They then compared the genomes with those from uncontaminated sites. They found the pollution-tolerant populations had each evolved similar adaptations that allowed them to live in toxic environments that would normally kill them.

Having adapted to their radically altered habitat, the little killifish were now up to eight thousand times more resistant to industrial pollutants than other fish. All this must have taken place in a few short decades, the authors surmised, since the most harmful pollutants (like dioxin and PCBs) had been released in the 1950s and 1960s. The killifish is not the only fish to have managed this feat; the Atlantic tomcod, a green-mottled bottom-feeder living in the polluted Hackensack River (a neighbor of the Passaic, which also empties into Newark Bay), is known to have evolved a gene that renders it immune to the harmful effects of PCBs.

The process almost certainly worked like this: of the massive killifish and tomcod populations that live on the eastern seaboard of America, a few individuals harbored genetic mutations that made them less sensitive to extreme toxicity. Most of the time this had little impact on their prospects, but should they live near a polluted site, they would find themselves at a distinct competitive advantage compared to their peers.

They breathed more freely, bred more successfully, and generally lived to pass on their mutation to their offspring, who did the

same. And thus a new pollutant-tolerant strain of killifish staggered into existence. You might call it *unnatural selection*. Scientists
call it "rapid evolution."

Rapid evolution of this sort was first observed in a very different species, in the UK: the peppered moth. A pale, delicately ornamented creature, the peppered moth leads an active nightlife—the
male whirring around in search of females, which lure their gentle-
man callers with an enticing pheromonal scent—but spend their
days resting quietly upon the trunks and branches of trees. For
centuries, this tactic worked very well; their cryptic, stippled wings
blend in so subtly with the pale, mint-frilled lichens that grow
upon the trees' bark as to render them near invisible.

But during the Industrial Revolution, when Blake's "dark, sa-
tanic mills" began belching their black smoke into the atmosphere,
everything changed. In Manchester, the epicenter of this new manufacturing activity,* fifty tons of industrial fallout settled over
each square mile of the city a year. The mills, the houses, the offices, the parks, the roads—all of them were coated in a thick
black residue. When it rained, the pollutants rained down with it:
sulfur dioxide, which kills lichen, and soot.

The trees in the surrounding region were now no longer wrapped
in English lace, but stripped to their barks and blackened. The

---

*Alexis de Tocqueville on Manchester in 1835: "From this filthy sewer pure gold flows."

peppered moths, resting quietly under branch, found themselves suddenly very visible indeed.

Good news for birds. Bad news for peppered moths. But not all peppered moths. In 1848, a Manchester lepidopterist named R. L. Edleston caught a rare dark form of the peppered moth, the first he had ever seen. The *"carbonaria"* form, as it came to be known, was ebony all over, except for two white dots at each temple. It carries a mutant gene that reverses its normal coloring. Usually this is a major disadvantage, and it dies out quickly. But now, with the trees soot-stained and lichen-free, it flourished. By 1864 the same collector found that this heretofore vanishingly rare creature outnumbered those of the original type. By 1895, *carbonaria* moths represented 98 percent of all peppered moths in Manchester.

You might be familiar with this story; it's one that's been told and retold in biology textbooks for decades, and made the researcher Bernard Kettlewell—who wrote the definitive paper on the subject in 1955—world-famous. The ascent of the *carbonaria* form so closely paralleled the rise in evolutionary thought, offering the first demonstration of Darwin's theory in action, that the peppered moth became an international poster child for evolution.

But this is no normal case of evolution. What makes the blackening of the moths and the toxin immunity of the fish so notable is the sheer speed with which they occurred. Natural selection is usually thought of as a painstakingly slow process. In fact, painstaking is the wrong word. Painstaking implies slow deliberate travel in a single direction. Evolution is based upon the opposite: sheer, random chance. Mutations might arise and be discarded thousands of times before an advantageous one arises, and even then it usually takes many thousands of generations—if ever—for it to become established.

In cases like the killifish and the tomcod and the peppered moth, we see a selective force so strong—an environment that has changed so much, so quickly—that it causes a bottleneck in the population. Industrial quantities of pollution are not the only cause of rapid evolution. Epidemics too have a bottleneck effect on a population—screening out all but the fittest and those with natural immunity. But humans have been the greatest single evolutionary factor now for centuries, and maybe even millennia.

Overfishing and overhunting have been fundamentally altering the gene pool, giving us smaller fish (all the better for slipping through the nets) and tuskless elephants. (In one South African national park, 98 percent of female elephants are now born without tusks, compared to a base rate of 2 to 6 percent.) Human predation in general is thought to have accelerated the rate of trait changes in other species by 300 percent. We too find ourselves in an arms race with pesticide-resistant insects, herbicide-resistant plants, drug-resistant viruses, and antibiotic-resistant bacteria.

And these rapid adaptations have taken unexpected forms. The very British habit of feeding songbirds in our gardens, just as one example, has produced in short order finches with longer beaks (all the better for using the bird feeder) and altered the (genetically stored) migratory routes of blackcaps. The London Underground, with its humid climate and pools of standing water under track, has spawned its own species of mosquito, whose isolation has rendered it no longer able to interbreed with its aboveground cousins. It—unlike its progenitors—now actively prefers the taste of human blood to that of birds.

Man-made climate change, plus its accompanying ocean acidification and the alterations to global ecology they entail, will too have untold evolutionary impact upon the world's flora and fauna. Not all species will be able to keep up with a planet that has been changing so quickly. In fact, there are reasons to suspect that the killifish in particular is almost uniquely well qualified to adapt. The winners in a rapidly changing world will be the creatures who, like the killifish, are highly numerous and genetically diverse. It's down to luck, in other words, and those with the most dice to roll might hope to be the luckiest.

Human industry has changed, and is continuing to change the world. Even if we were all to be wiped out tomorrow—factories falling silent; generators shuddering to a halt; cargo ships drifting and colliding, sinking to the seabed, sending sediments billowing—we have set in motion evolutionary forces that will continue to act upon the genetic makeup of almost every other species alive on this Earth. They shape-shift and metamorphose, transmute and adapt, in ways that we cannot anticipate and certainly cannot control. They want to live, if they can.

On the shore of the Arthur Kill, a red sun is rising. To the west, beyond the wrecker's yard, the smooth pale gasometers of

Chemical Lane rise to meet the sky, color-shifting with the sky: rose to lilac, primrose to forget-me-not.

All is still, barring the low scuffling of the breeze through salt hay. I step out onto the foreshore and sink immediately into a thick, oozing mud that comes over my ankles and sucks at my boots. As I do so, I disturb a flight of hooded black birds, cormorants perhaps, that lift and skim away over the surface of the water. Through the rusted spires of the sunken ships, I see as if through smoke the refinery towers rising as a retro-futuristic vision along the skyline. A vast gray container ship glides silently through the haze between us with a smooth and alien momentum.

Here, in the museum of abandoned industry, that anything continues to survive in the toxic water at my feet strikes me as beyond fortunate—an act of divine providence, a gesture of forgiveness on a grand scale. Among the rushes lie scattered the ragged discards of civilization—sodden tissues and plastic bottles and shopping bags—and irregular puddles bearing the iridescent luster of petroleum; a reminder of the oil spills that curse this waterway.*

The whole scene—my scrap of wetland, the rusting ships—stands in the shadow of an immense, grassed-over mound: an invisibility cloak cast across what was once the world's largest garbage

---

*In 1979, 230,000 gallons of crude oil were spilled near here, 567,000 gallons of heating oil in 1990, and numerous others of a smaller scale. Only a few weeks after my visit, an estimated 100,000 gallons of diesel were spilled yet again into the Arthur Kill. Oiled birds and "tar balls"—hard, weathered pebbles of oil—were washed up on the shore.

dump and remains one of the largest man-made structures in history. The Fresh Kills Landfill, the source of the "syringe tide" of 1988, when hypodermic needles and vials of HIV-positive blood washed up on the Jersey Shore. Now, like the dioxin-soaked remains of the Diamond Alkali plant, it lies sealed away behind layers of clay and concrete.

These accursed tombs, sealed up the best we know how, are our culture's equivalent of the Valley of the Kings. These are the monuments we have left for future civilizations to remember us by, and the PCBs and dioxins and the myriad other persistent organic pollutants that hide inside them will live on, essentially, forever. Certainly, many of them will last longer than a man-made catacomb will remain airtight: a new curse of the pharaohs, waiting to escape.

But the mutant fish and carcinogenic crabs of the lower Passaic and Newark Bay are evidence that the dystopian future is already here. It has started. Though industrial contaminants like dioxins and PCBs have deeply impacted the environment in sites like this around the world, the killifish and the tomcod show us that life—and the evolutionary forces that shape it—is dynamic. It can, under certain circumstances, bounce back.

If dioxin is the curse, then this is the blessing. Nature has—or, rather, some of her constituent members have—the capacity to survive in conditions that would once have killed them, the ability to adapt to a befouled and ruinous world, and even thrive there. Perhaps only a tiny proportion of the glorious array of life that exists today will have such remarkable superpowers hiding in their genes. The thought of the others, the sheer scale of what might be lost, what is already lost, is dizzying. But it offers a tiny, hairline crack, through which light glints.

In the UK, parliamentarians were finally convinced to take action on the air pollution problem that produced the melanic moths when the Great Smog of 1952—a foul "pea souper"— killed an estimated four thousand and made working in Westminster impossible. The new Clean Air Acts that followed heralded in an era of improved air quality and huge reductions in the release of sulfur dioxide all across the country. Over time, lichens returned to the trees. And with them came changes in the selective pressures on moths. In West Kirby, near Liverpool, the geneticist and lepidopterist Sir Cyril Clarke and his wife, Frieda, spent thirty years watching the proportion of the dark *carbonaria* form of the peppered moth, associated with poor air quality, fall off a cliff as the original, pale, lichen-like variety returned with a vengeance: 93 percent of peppered moths were black in 1959; 53 percent in 1985. By 1989, there were less than a third.

Nature reacts and responds, reacts and responds. Here, on the Arthur Kill, the ruins of past industry lie rusting in the bay. Tainted water laps gently at the hulks of the sunken ships. The worst is over, the leaking waste pipes capped. But this time, how long will the smog take to dissipate? I run the sums in my head: if dioxin takes, say, a century to half-degrade, and if it can still be harmful at parts-per-trillion levels . . . The haze is dark and dense.

Beyond the ships rises the domed peak of Fresh Kills, the landfill covered over and now rippling with waist-high, heavy-headed grass. Light is flooding through the sky. Birds fly overhead. It is the dawn of a new day.

❦ 8 ❦

FORBIDDEN FOREST

Zone Rouge, Verdun, France

After weeks of snow and sleet and storm, the morning of February 21, 1916, dawned fresh and clear and hard. In the hills above Verdun, on the Western Front, two armies faced off across a wooded hinterland. To the west, hot breath rose and condensed in a thin pall over French troops as they crouched in their hastily dug foxholes.

All was still. Boot-trodden puddles in the foot of the trenches had set overnight into thin panes of mud-stained glass. The soldiers were dog-tired, worn out already by anticipation of an attack they knew would be coming. They were not—perhaps never could be—prepared. But after days of waiting, they were almost impatient for it to begin. And now, as a pale wash soaked into the sky, they knew it would be soon.

Across no man's land, the German artillery crews too were awake, loading long-range artillery guns, just as they had done every dawn for the last nine mornings. When the order finally came, at a few minutes past seven, they were ready.

Within moments, so many heavy guns had opened fire that, to the French, the German front appeared a wall of leaping flames. The ground, wracked by impact after impact, began to quake in a rolling, cataclysmic tremor without end. Shells came raining down in every direction, shattering all that they came into contact with, sweeping the landscape like a hose.

Nine hours of this: an incessant rain of death without let up. Afterward, when finally the weaponry had fallen silent, the Germans sent pilots up to reconnoiter behind enemy lines. They returned, shaken to the core. It's done, they reported. It's over. There's nothing left alive.

But it was not done. This was merely the opening bombardment in a battle that would rage through the days and weeks and months to come. Soon this hellfire would become normality: the constant storm of artillery, a thunder reverberating in the chests and the lungs; the earth a pulverized mass of mud and clay and jellified blood and fragments of bone, flesh, and shrapnel pounded into paste; the corpses piling mutilated in the trenches, entrails spilling—buried in a tide of filth only to be blown bare once more. Men struggling with rifles in mud-soaked uniforms, on their knees in filthy water, stepping on the bodies of the dead. Gibbering men, sent mad by fear; silent men, sinking, shell-shocked, to the ground, emptied out by the horror of it; grim-faced men, following orders, fighting.

That year, rain fell almost constantly: the dull and dreadful rhythm of it came pounding off steel helmets. Steam rose from shoulders, seeped through uniforms, trickled between toes. Water was everywhere, yet they had nothing to drink. Men sucked from the filthy ponds that filled the shell holes, drank their own urine, licked moisture from the walls of fortifications. All this, for three hundred days straight.

It was, by a number of measures, the very worst place in the world. This, the Battle of Verdun, was the very origin of ultra-violence: the first time the world had seen killing on an industrial scale. Though other battles of the First World War would later claim more lives in total, Verdun—where shelling raged all through the summer, and then through a bitter, frozen winter—claimed the highest proportion of its combatants' lives, and in the most concentrated area. An estimated 300,000 men died here, and another 450,000 were gassed or wounded, over less than eight square miles. It remains the longest battle the world has ever seen.

When, finally, it came to an end, survivors emerged as if from a nightmare and looked upon an annihilated landscape, a barren wilderness that stretched off in every direction without landmark. Perhaps forty million shells had been fired in these hills, more than six shells to every ten square feet of ground. They left behind a churned-up sea which pitched and rolled with unseen currents. Bleaching bones and the broken limbs of rifles protruded from the wave. Between 1914 and 1918, the soil had undergone the equivalent of ten thousand years of natural erosion.

All that remained of the ancient forests of Spincourt and the nine *villages detruits*—the "villages that died for France"—were the shards of trees poking haphazardly as tombstones; the rubble of stone foundations; the knotted strands of barbed wire. A dead zone, a flayed and featureless creature stretching off in all directions.

But it was not dead, not truly. In the summer of 1917, a fur of scarlet poppies grew up, softening the gaunt hollows of this broken land. A sign, the soldiers thought, of hope. A glimpse of a thought that life might yet continue, no matter how unlikely that must have seemed.

The battle was over, but the war was not yet won. The stench of rotting corpses hung in the air.

Two years later, all along the Western Front, French authorities were taking stock. From the border with Belgium at Lille to the border with Switzerland near Strasbourg, this most brutal of wars had torn a rupture through the land: it was ripped, cratered, pitted, charred by a billion artillery shells fired over four years. "Where there are no dead," wrote Henri Barbusse, "the earth itself is corpselike." A Frankenstein landscape, stitched and stapled together, which harbored in its flesh millions of tons of unexploded munitions and chemical weapons enough to kill an army, all over again.

Bodies too, of course. Of those men who died at Verdun, only half were recovered and identified. The rest were left so mutilated as to be beyond recognition, or were found in parts, or had vanished into the mud. It sucked them under, like quicksand, then sealed again over the top.

The country was still in shock, but thoughts turned as they must to the work of reconstruction. Economically hobbled by war, and awed by the scale of the problem facing them, the French authorities developed a triage system: they surveyed the *régions dévastées* and drew up a series of maps which charted areas believed to be devastated beyond repair. In total, more than four hundred and sixty square miles were classified in this way, shaded with a red pencil and declared no-go areas.

Over the following decades, this total was much reduced. The battlefields of the Somme and Ypres (across the border, in Belgium) lie in valuable agricultural areas, and were later returned to farmland—albeit as farmland where, every year, farmers continue to unearth more shells and rusting canisters as they work their way to the surface like rocks, so many that the plowing season has become known as the "Iron Harvest." But near Verdun, where the land was choppier, steeper, more remote, and the damage total—a "biological desert," as the French botanist Georges H. Parent described it—much remains off limits.

After a period of uncertainty, it was decided to plant a shroud of trees over the war zone—a living sarcophagus that might stabilize the soil and contain the terrors within for a generation or more. A forest of forgetfulness. Where the land was worst affected—soil stripped almost to the bedrock—they planted black pines, one of the only hardy species that could thrive there. So it remains today. They call it the Zone Rouge.

And so for a hundred years a forest grew up across the land, tall and dark and impenetrable, whose undergrowth curled and snarled into a thicket of bramble and blackthorn. This was a forbidden forest, a tangled briarwood that protected not sleeping princesses in their castles, but the horrors of war that still lay dormant under a thin skin of earth.

Signs along the perimeters warned visitors of the dangers of trespassing, of straying from designated pathways, of stumbling upon the lethal remnants of a war that, though long over, nevertheless continues to claim more victims almost every year. So much ordnance remains that the casual observer, treading quietly upon a permitted path, might still stumble upon the dry flakes of shrapnel among the leaf litter, the rotting branch of a rifle, a polished pebble of lead shot, as if the men had simply set down their weapons, laid down on the earth, and turned into trees. Disregard the warnings, step beyond the safe passages, and risk stepping on the unknown, unblown bomb.

Here and there, a few aging oaks remain, two centuries old, improbable veterans of the onslaught. They are strung with copper cables, weighted with steel bars and electrical insulators, dating from their past service as observation towers, holders of power lines, church spires in a time of unlandmarked ungodliness. The ground between their roots stands in stiff peaks, slashed through with trenches still—semi-healed, scarified—gouged with shell holes and scraped bare by wild boar searching for bones. Here and there, amid the elderberry and the ferns, nudge the moss-softened forms of bunkers and the black mouths of dugouts fallen derelict. (Inside: the papery sounds of bats moving around in the dark. Pale fingers of stalactites drip like icing to the floor.)

Not all is left forgotten. Near Fort Douaumont: a vast ossuary. *Ossuary*: from the Latin, *ossuārius*, an urn for the bones of the dead. A hall of remembrance built to hold the skeletal remains of 130,000 men. Approximately. Windows into the bowels of the building reveal the chaos within, the impossibility of ever really knowing for sure. Skulls lie heaped upon skulls. The beams of thighbones stacked like timber at the side of the road. Balls and sockets, scapulas, pelvic girdles—the joints and joists of humans broken into their constituent parts. Unearthed in disorder, they have been stored according to the sector of the battlefield in which they were found.

Inscribed upon the walls in the chapel above, the names of the missing. There, with the echoing whispers of a hundred strangers merging into a chorus of grief, their faces lit with an amber glow, it seems too much to bear. And then one departs through the front door and looks up to the vastness of the sky above—so clear and cloudless as to seem black, to hurt the eyes—and the stone tower standing sentinel overhead, ascending as if to the heavens and holding within it the *lanterne des morts* that sweeps over the massed graves like a lighthouse in the night. And in that moment, a half-dozen swallows leap from the belfry to float and swell through the air: looping, soaring, sailing free, and the soul lifts.

And again, later, in the dark woods, while shadowing a winding trail through the foundations of the lost village Vaux-devant-

Damloup—where offerings lie in tribute to the spirits of the ancestors, whose buildings' spectral forms seem to shimmer between the columns of the trees, and—the sharp edges of the stones are dressed in felted green, and the water troughs made planters of their own accord. Where the scarlet berries burn like embers under lacquered leaves, and pale butterflies somersault through the shafting light.

Here, life lurks in the folds and wrinkles: fifteen species of fern jostle for space in shadows; thyme scrambles over dry rock, searching for purchase where the trees found none; ponds form in the craters, in which skulk newts and yellow-bellied toads. Rare orchids arise along the edgelands. Songbirds lift their voice in praise. Vegetation breathes. Here one feels at once the astonishing fortune of living in such a vast and endlessly forgiving world; the beauty of it; the blessed relief. If there is a God, he may yet be revealed as a merciful one.

But not five miles from here is a place where the trees never did grow back: a clearing in the woods, where the oaks and hornbeams part to reveal a small round pool of what appears to be gray gravel, or tar, or ash. A swatch of ground where nothing will grow.

The secret to this sterile wound lies also in the decisions made after the war. At armistice, millions of unused shells lay piled up ready to be fired. It wasn't clear what should be done with these surplus weapons of mass destruction. At Verdun, the decision was made to recover what ammunition they could at the military camps, but to gather chemical weapons—two hundred thousand of them—at a farm near Gremilly. Here was mustered an array of the most unpleasant hexes one man can cast upon another: mustard gas, tear gas, phosgene—whose pleasant odor of freshly mown hay belies its deadly consequences—the sneezing gas diphenylchloroarsine, the garlic-scented vomiting agent diphenylcyanoarsine. Then, finally, in 1928, they dug trenches as if for a mass grave, piled in the canisters, and set them ablaze. Hence the name: la Place à Gaz.

The clouds of arsine gases produced during the burn poisoned the land and left it bare. It appears like tundra, or melted tarmac: waste ground of the very purest kind. At the center, a tar-like ash lies dark and bare, its surface ruffled like choppy waters. Lichen and moss creep around its very edges. Farther back, hair strands of thin grass fall in partings beyond. Nothing else. Even as the forest grew and stretched around it; in the Place à Gaz it was always winter, never spring.

Though the name persisted, its origins passed into obscurity— lost not so much to the mists of time as to a collective amnesia. Understandable, really, when you think of what the local populace had gone through. As the years went on, huntsmen and foresters stumbled upon the clearing afresh and mistook this cursed ground— lurking with a hidden malevolence between the trees—for a pretty, dappled glade where one might stop to catch one's breath and eat a bite of lunch in the sun.

At some point, someone enterprising built a small hut on the edge of the clearing, a single-roomed cabin with a corrugated roof and a tiny chimney and a window that looked out upon its own private rock garden. To the unsuspecting eye, it must have seemed a peaceful sort of place. Compared to the dark forest beyond, which in winter sagged with dank vegetation, and where wild boar and roaring elk roamed, it must have felt safe.

It wasn't. In 2007, the German scientists Tobias Bausinger, Eric Bonnaire, and Johannes Preuß identified this barren patch of ground from the historical records, and undertook a chemical analysis of its soil. In places, they discovered, 17 percent of the soil's weight was made up of arsenic. And plenty more of what biologists call heavy metals: up to 13 percent zinc, 2.6 percent lead. With a dawning horror, the huntsmen realized that for years they had been walking on, sitting on, eating on a carpet made of poison.

The reason it stayed bare was that nothing could grow there. Or, almost nothing.

Many of the so-called heavy metals* are essential to the fundamental processes of life, but in quantity they become toxic. When plants come into contact with metal-tainted soil, strange things can happen.

In the 1950s, the Russian naturalist N. G. Nesvetaylova discovered that it was possible to turn poppies different colors by the

---

*The term *heavy metal* is used by scientists in a number of different contexts. Even laying aside those with extracurricular enthusiasms for Slayer or Judas Priest, we might find the term used to refer to metallic elements of particular densities (>5 g/cm3), atomic weights (>40), or positions in the periodic table (within the transition metals). More generally, however, it has come to be used as a catchall term for metals considered toxic to plant and animal life, including cobalt, copper, iron, nickel, and zinc.

adding of various metal salts to compost: zinc compounds produced flowers of lemon yellow, for example, whereas boron turned their leaves dark green. Copper, on the other hand, produced pale, blueish, "dove-colored" leaves. (In this way a gardener with fairy godmother aspirations might sprinkle manganese on the soil beneath an almond tree, to turn their flowers' corollae from white to pink; aluminum sulfate over the roots of a hydrangea will turn its cotton-candy heads mauve, then indigo, then baby blue.) And there was a combinatory, witches' brew aspect to the process: two or more salts added together, like a tincture, and the flowers would take on unexpected new shades, wholly different from those seen when the metals were added separately.

Large-mouth poppies (*Papaver macrostomum*), common to the Middle East and Kashmir, develop double-decker petals when growing in high-zinc soils, while the ladybird poppy (*Papaver commutatum*) of the Caucasus alter the pattern of their spots in response to copper-molybdenum. In the areas of greatest mineralization, their dark spots elongate until they meet at the center to form a cross—*x* marks the spot—a signpost to the contents of the underworld.

Plants growing in the vicinity of manganese might boom obscenely in size, reaching gigantic proportions with luxuriant greenery. Copper sulfate or chromite will produce dwarfs. Symptoms such as these have been used successfully by prospectors across the globe for centuries as "bioindicators" of minerals in the soil below. As explorers once scanned their environment for willow or cottonwood to find water in the desert, prospectors raked the landscape for plants displaying chlorosis, the floral equivalent of anemia, which shows up as a bleaching or fading of the leaves, except along their darker veins, which stand in dramatic silhouette.

Better yet, they might spot plants whose presence alone signaled valuable metals. Early Scandinavian miners, for example, were guided to their target by the *kobberblomst* (copper flower) and the *kisplante* (pyrite plant),* tiny pink-flowering campion whose delicacy belies an extreme hardiness, an ability to thrive where sometimes no other species can.

By the sixth century AD, the Imperial Chinese were already sensitive to the potential of metal-loving plants as a means of prospecting, producing detailed manuals listing different species and their mineral familiars, and the symptomatology associated with specific metals, their instructions sounding with the prosody of incantation, of summoning spells. ("If leaves ... are green, and the stalks red, much lead will be found below ...")

In such way, the expert geobotanist can glean a great deal of complex information from flora. In the Copperbelt Province of Zambia, for example, at least twenty-seven flowers live almost exclusively in soils tainted with copper and cobalt: the thicker the pelt of these flowers, the greater the degree of mineralization. Similarly, in the Alps one might learn to predict the presence and concentration of zinc by the depth of color of the flowers of the tiny, lemon-yellow calamine violet. In Australia, two plants—a type of flowering pea (*Tephrosia*) and the paper-petaled herb

---

Lychnis alpine.

Polycarpaea spirostylis—come together to form a map with contours: the *Tephrosia* tracing the outer edges of copper deposits, but giving way to the flag-waving *Polycarpaea* wherever the copper reaches more than 2,000 parts per million.

Some regions are so profoundly impacted by the presence of concentrated metal ores that even these rare "metallophytes" cannot get by. Such areas, denuded of vegetation, might appear as a sickly pasture that pockmarks otherwise lush, forested areas. Platinum has been found under bald spots in the Urals and South Africa; in Russia, boron. Folk stories spring up around such oddities: one such spot in North Carolina is known as the "devil's stomping ground"; the source of its barrenness is yet to be determined, and may predate human presence in the region. But for heavy metal contamination to be so extreme as to render a place barren dirt is rare.

The Place à Gaz, that clearing in the forest near Verdun, is such a place.

It's hard to get directions to a place like the Place à Gaz. People hesitate to share information about arsenic deposits sitting open to the air. So I arrived in Verdun with no clear sense of how I was going to find it.

Over a late night in a cut-price French motel room—textured orange walls, buzzing fluorescent bars, carpet tiles, outdoor pool flanked by aging plastic lounge chairs—I sat hunched at my laptop scouring satellite images of the forests of the Zone Rouge, guessing at distances, looking for a sign. Finally, I alighted on what appeared from above, in certain seasons, as a mauve thumbprint in the woods, not far from the farm where weapons were stockpiled. I knew immediately I had found it.

The next morning, I drove slowly along the track, watching the GPS. When I drew near, I pulled in, reversing self-consciously to park among the undergrowth at the edge of the trees, and walked quickly and directly into the woods. Within a few moments I came up against a military-style enclosure: a fence of wire mesh of perhaps eight feet in height, further buffered by a concertina of razor wire around the base. It was difficult even to get close enough to press my face to the fence. A laminated letter, nailed to the trunk of a nearby tree, stated that *l'accès a l'intérieur de la zone* was *interdit*. Access prohibited. They weren't joking.

All I could see from there were the bare backs of trees, their attention caught by something farther in. I craned to see, and followed the fence line deeper into the woods. Beyond it, I saw the vegetation part. Inside this razor-briar lay the clearing: open and peaceful, tiger-striped with shade. Pale butterflies flitted and tumbled through the tall grasses at the edge of what appeared to be a still, slate-gray pool. Tiny insects drifted like motes of dust through shafts of light. The tiny cabin perched at its edge, door swung open invitingly. A fairy-tale dell sealed behind a heavy-duty, military-style compound. Its innocuous appearance jarred with the scale of the defenses surrounding it; my senses buzzed with the disconcerting sense of danger camouflaged in plain sight. What it most brought to mind—powerfully, ominously—was Andrei Tarkovsky's great masterpiece *Stalker* (1979)—a film set in a dystopian near future, in which a Soviet man known as the "stalker" guides two strangers into a mysterious, tightly guarded exclusion zone.

In *Stalker*, the three men evade the guards and find "the Zone" to be a region of peaceful, verdant beauty. The stalker leads them onward, slinking through overgrown meadows knee-deep with thatch, skirting ruined buildings overcome with vegetation, wading through rivers and overflows—and as he does he constantly tests the ground ahead of him, throwing metal bolts every few paces as he searches for the invisible, unknowable dangers that lurk unseen throughout this sylvan glade.

Myself, I began a slow circumnavigation of the compound, which took me around the rear of the deserted woodsman's hut, where a sagging lean-to sheltered slowly decomposing logs stacked neatly at its back. Shards of a white sign lay shattered on the ground:

IN   TERD   IT

The grass was golden, wisp-thin like spun sugar, shimmering in the breeze. At the center of the clearing that pond of arsenate ash lay still. Mosses and tufts of grass poked through at its margins as if through shallows, the center clear and barren. Climbers wound their way over the fence; muscled ivy shinned up the trees at the edge of the clearing. Low branches leaned in over the gap. Vegetation pressed in on all sides, but could not enter. Cursed ground.

Tarkovsky believed that *Stalker* was cursed too. The production was troubled from the beginning. An earthquake ended plans to film in Tajikistan; then, having moved to Estonia, a year's work was destroyed while being developed. The following year, the re-run shoot was delayed, impossibly, by summer snowfall.

And though it was finally completed in 1979—to immediate acclaim—more misfortune would follow. Anatoli Solonitsyn, the star, died three years later, aged forty-seven. Tarkovsky, at fifty-four, four years after that. In 1998, Tarkovsky's wife and assistant director, Larisa Tarkovskaya, passed away—all from the same rare form of cancer. Crew attributed it to long days downwind of a belching chemical plant; to their days spent wading, knee-deep, through the toxic froth of the Jägala river during filming.

A rustle came from the undergrowth. A lone doe emerged, trailing her muzzle almost thoughtfully through the long grass, and, with a careful delicacy, picked her way over a babbling brook I knew to be laden with poison. I stepped forward into the light and she froze, eyes on mine, for a long and meaningful moment. Then, she turned and fled into the forbidden forest.

Those plants that grow as a wan halo around the poison ashes: it was those I was here to see.

On first sight, they seemed disarmingly familiar: the haze of what, at home, we call tufted grass and the Americans call "velvet" grass for its peach-fuzz leaves—common to marshland, verges, neglected waste ground—and, hidden beneath, like an underfur, the powdered goblet lichen *Cladonia fimbriata.* Neither are exotic species. But plants like these are specially adapted to survive in what would be otherwise be a dangerous environment. They limit their intake of the metals, preventing a build-up to toxic levels in their bodies. Quite simple.

Their neighbor, though, a soft and feathery moss known as *Pohlia nutans* ("nodding thread moss," after their tiny, many-headed fronds), uses a more complex strategy: rather than close itself off to the metals in the soil, it throws open the doors, transporting metal salts upward into its limbs and stashing them away. Magpie plants of this kind are known as "hyperaccumulators," and it's not totally understood why they do this. It may be a form of self-defense: making themselves bitter herbs, to deter grazing animals.

The effect, though, can be extraordinary. For example: the *Pycnandra acuminata* is a silver sylph of a tree that grows in the misty rainforests of New Caledonia. When cut with a knife, it bleeds a spectacular latex sap the color of verdigris, containing up to 26 percent nickel. In post-industrial mining regions of Wales, lichens soak up iron or copper from the rocks they grow on, turning rust-orange or turquoise in the process—the spattered paint of an artists' studio—and rendering the metals insoluble, and therefore harmless.

Though metallophyte species like these have evolved naturally, finding toeholds in outcrops of metal ores and at sites like the Tantramar "copper swamp" of Canada's New Brunswick, they are now much more likely to be found in human-impacted ones. Mine tailings, spoil heaps, slag tips, post-industrial sites of many kinds—and post-conflict ones too, like the Place à Gaz. There has been an exponential growth in land despoiled by heavy metals over recent decades. Globally, more than five million such contaminated sites have been reported; more than three hundred thousand square miles of contaminated soil in China alone.

Knowledge and appreciation of such plants has grown in recent years, but their love of despoiled landscapes gives conservationists pause when it comes to protecting rare and unusual species. Near Swansea, in southern Wales, I once visited the former site of "Copperopolis"—an enormous complex of smelting furnaces built during the seventeenth and eighteenth centuries, which left the lower Swansea valley a lunar landscape besmirched with lead, chromium, and copper in the years after the industry's collapse. This scrappy wasteland, however, had recently been designated a Site of Special Scientific Interest thanks to its unusual assemblage of metallophyte plants and lichens—known as "calaminarian grassland" after the zinc ore—and brought under new environmental protections.

Disturbance, though, turned out to have been very good for the area; despite the best of intentions, an early "remediation scheme" which removed or capped metal-polluted soil had reduced the habitat of rare plants like the star-flowered spring sandwort, and conservationists were now considering radical, counterintuitive management methods—such as scraping up the topsoil to *retoxify* the ground.

Because of their strange and beguiling qualities, metal hyperaccumulators—of which there are known to be around five hundred—are of enormous scientific interest. Thanks to their thirst for otherwise toxic materials, they have great potential as tools in the recovery of highly polluted sites. By sucking heavy metals from the earth and hoarding or redistributing them, they might prepare the ground for other, more sensitive organisms. In this way, nature begins to heal over her scars.

Already I could see this process at work. In the Place à Gaz, the bare surface of chemical ash had clearly declined since the German scientists' 2007 study, and perhaps even since a French follow-up paper in 2016, which noted with relief "progressive revegetation of the site." Whatever these plants were doing—and particularly the nodding thread moss—was slowly turning the chemical burn in the landscape into a habitable place to grow.

A field of study, phytoremediation, has grown up around hyperaccumulating plants. It seeks to harness their surreal kind of superpowers for the greater good. Other species include the brake fern, which removes arsenic from the soil and stores it in its fronds (and is being tested as a natural filter for contaminated water in Bangladesh, following a decades-long arsenic-poisoning crisis), and sunflowers, which accumulate a wide range of heavy metals and are grown on sites of former mines and smelters in Australia.

It's a slow process; the plants must grow and then be harvested—and their bodies, now containing high concentrations of the heavy metals, disposed of carefully—but it can be faster and certainly less environmentally damaging than current clean-up methods: excavation and reburial under a concrete cap. And if the plants hold enough metal in their flesh—at levels of over 150 parts per million—another intriguing possibility is raised: the plants themselves might be treated as a sort of organic ore—dried, burned, and the metals extracted from their ash for reuse. In this way, a farmer might grow and harvest a "crop of nickel"—or cobalt, or even gold—and make a larger profit than he does now from barley or wheat.

It was commonly believed by alchemists in the sixteenth and seventeenth centuries that metals and other minerals were merely a lower form of life, which grew from seeds in the bowels of the earth, ripened and replenished themselves. As John Webster wrote, in his 1671 *Metallographia: or, an history of metals,* "that Metals do grow even like other Vegetables is manifeſt from divers examples." He quoted Peter Martyr, counselor to Charles V, Holy Roman Emperor: "the Vein of gold is a living Tree, and that the ſame by all ways ſpreadeth and ſpringeth from the root . . . putteth forth branches . . . ſheweth forth certain beautiful colors in the ſtead of flowers, round ſtones of golden Earth in the ſtead of fruits, and thin plates in ſtead of leaves."

Now, with the prospect of stems threaded through with copper and leaves of gold, fresh stumps spilling with a silvern sap, flowers dripping their nectars of nickel; now that we find ourselves in a world iced with arsenic and soaked in lead, and it all the work of our own hands, the notion feel less ignorance than foresight. Wrote Milton:

And all amid them stood the Tree of Life,
High eminent, blooming ambrosial fruit
Of vegetable gold; and next to life,
Our death, the Tree of Knowledge

Toward the rear of the enclosure, at the farthest point from the road, a hole has been dug under the fence. It might have been made by a badger, even a dog. Not a large hole. But then, I am not a large person.

I shimmy under it and into the clearing. I clamp my mouth shut, screwing up my eyes against arsenic-laced soil, and pull myself on my back through the hole, carefully positioning my hands between the razor barbs and unhooking my shirt as I get snagged.

On the other side, I stagger to my feet, step from the tussocky shore onto the pool of barren gray ashes, the cursed ground. It shifts, soft as sand, though the edges are solidifying under plates of moss. Feeling unsteady, I step out again into a shaft of light that cuts through the trees and stand illuminated in my circle of scorched earth. All around—but held back, as if by some invisible force field—the vegetation of the forest shines green and verdant and vivid, intoxicatingly bright.

During the Victorian era, there was a fashion for a certain shade of dye: a vibrant emerald whose invention sparked a craze for green dresses, green wallpaper, artificial greenery of all kinds. The catch was, it was highly toxic, a mix of copper and arsenic trioxide. Factory girls died of it, foaming green at the mouth; their eyes and fingernails, stomach, lungs, all green. A woman in a ballgown of "Scheele's green" carried enough arsenic in her skirts to kill everyone in any room she ever cared to enter.

Four or five grains of arsenic is enough to fell a grown man; over the course of an evening, sixty grains might powder off from a dress and onto the floor. How many grains of arsenic, I wonder, might be found in the ash of two hundred thousand chemical weapons? (How many, I thought—unable to stop myself—have I just brushed from my jeans?) Still, even after the truth was known, people couldn't resist the charms of this unearthly green. The same urge, I assume, that has driven me under the fence. I shine the poisoned apple. Take a bite.

Inside the enclosure, I feel a sudden surge of adrenaline, even panic. A breeze riffles through the long grass. Birds sing in the branches behind the fence. I swing around, eyeing the fence. There is no easy way back out. I catch my nerve, turn back to look at the darkened room. My feet sink into the soft gravel of war, the precipitate of man's self-annihilating impulse. A circle of barren ground left like a fingerprint at a crime scene, evidence of a war unprecedented in its scale and destruction, in its reckless devastation.

In *Stalker*, visitors are drawn into the Zone by the promise of a mysterious room at its heart: all those who enter will find their greatest wish come true. But the stalker, though he has guided countless others to its threshold, has never himself entered. The implication is, I think, that despite the risks—despite the lurking, unseen horrors dressed in fairy-tale clothing—this treacherous zone offers those who pass there something more valuable than earthly riches. In a corrupted world, it offers hope.

I take another step toward the room at the heart of this small zone. The woodsman's hut, and the dark, deserted space inside. The door is hanging open.

9

ALIEN INVASION

Amani, Tanzania

The rain is not long over. Vapor rises from the vegetation and swirls between the trees. There is a faint drumming of drops falling from a great height upon soft earth. The air is warm and wet as bathwater, and thick with the aroma of the eucalyptus. Spotted gums stand in rows along the hillside, shedding layers like veils upon the ground, droplets glistening on their hot silver skin. I breathe it in, finding comfort in familiarity. *I know this place*, I think.

But something's off. It looks and smells like the Australian bush, but the soundtrack is wrong. My ears search for that glitchy dial-up garble of the Australian magpie, the bellbird's sonar pulse, and find nothing that matches. Instead, over a drone of cicadas, a thin avian voice whistles a melodic falling scale. I hear the honk of something like geese, the soft whoop of what can only be a monkey.

Soon I pass, as if into another room, into China. Bamboo sweeps the path like a curtain, loose-weave, the color of straw. Here and there, outcrops of its giant variety burst from the ground in clumps, tree creatures with their knuckled limbs outstretched. I walk on. The bamboo fades out, is replaced by a South American palette: the monkey puzzle,* with its huge spilling limbs and spear-tip scales; the fronds of the Spanish cedar,** the pale petal trumpets of the cinchona. Then oil palms, red fruits bunched and rotting on the branch, mark a detour into west Africa.

In truth, we are in none of these places. Before my feet, striped lizards run pell-mell. Huge armor-plated millipedes cruise the leaf litter. Farther down, close to where the river runs clear, a red crab skitters from the undergrowth and between my feet before vanishing into vegetation.

This is Amani, high in the Usambara Mountains of Tanzania, but I could be anywhere at all. This abandoned botanical garden offers a cautionary tale of the dangers of transporting species around the world. But Amani may too show us a glimpse of something else: of the surprising ability of species to rub along with one another, even if they ought never to have met. Their success in finding novel ways of coexisting offers us hope that in Amani, and in so many other sites around the world, ecosystems may be more flexible than we have assumed.

Araucaria araucana.
**Cedrela odorata.*

The Amani Imperial Biological-Agricultural Institute was established by German colonial powers in 1902. Trial plantations of more than six hundred species of trees and woody shrubs were established in an experimental arboretum, along with laboratories and a botanical garden containing two thousand further species. The hope was to identify crops—timber, oil, rubber, fiber, fruit, spices, coffee—that might adapt to the clay soils and tropical climate of what was then German East Africa. Under the direction of Professor Albrecht Zimmerman, Amani grew into the largest and most important herbarium in Africa.

Still, Amani's activities were not unusual. Imperial powers of that era transplanted crops between continents, shipped livestock around the world, and clear-cut native forests to make room for it all—and all this on almost unimaginably vast scales. (Australia, for example, was settled in 1788; by 1890 it supported at least a hundred million sheep.)

As well as through agriculture, the transporting of exotic species furthered imperial ambitions in many unexpected ways. In the nineteenth century, for example, hundreds of thousands of giant tortoises were removed from the Galápagos after it was discovered they could survive in the hold of seagoing vessels for more than a year without food or water. Their flesh was delicious ("no animal can possibly afford a more whole, luscious and delicate food," one sailor panted) and, better yet, they carried in a special "bladder" several liters of clean, potable water. How convenient, for an ocean-going empire. Quinine, an extract of the Andean cinchona tree, was the only known cure for malaria for three centuries. Without it, the colonization of the tropics could have foundered in fits of fever.

Sometimes colonists took plants and animals with them simply for company, or because they looked pretty, or gave them a sense of familiarity in strange lands. It was a popular colonial pastime. In 1890, an "acclimatization society" in New York released into Central Park examples of every avian species mentioned in the works of Shakespeare.

> The ousel cock, so black of hue,
> With orange-tawny bill,
> The throstle with his note so true,
> The wren with little quill.
> The finch, the sparrow, and the lark . . .

Most disappeared into the trees and were never seen again, but the sixty starlings ("I'll have a starling shall be taught to speak/ Nothing but 'Mortimer'") flourished. The American starling now numbers approximately two hundred million.

In Amani, the Germans too wanted to feel at home. They cleared the trees from the mountaintops to build their villas, and planted front gardens with neat lawns of kikuyu grass, imported from Kenya. They brought in dairy cattle to keep the grass trimmed and to fertilize it with their dung. They built a small train line, and for the stationmaster a Swiss-style chalet in dark-stained timber, with carved balustrades. All of which combined to create the disconcerting effect of an Alpine meadow—but for the red-rust earth that shone through wherever the grass grew worn.

But Amani wouldn't stay in German hands for long. War broke out in 1914, and in 1916 a small contingent of British forces—led by a local missionary—marched over the hilltops and took the research station. To the Brits, Amani was a great windfall. They switched the focus of the research there from botany to medical research: to curing malaria, specifically. Good work, if you could get it: an ivory tower–cum–hilltop retreat, where the scientists lived privileged lives in airy colonial villas with whirring fans and gleaming mahogany floorboards. They installed a library, a post office, and two social clubs—one for the white scientists, and another for their African lab assistants.

In the evenings the white scientists drank gin and tonics on the terrace, or played tennis, or paddled amid fireflies on the boating pond. "On a few clear days before the rains," recalled one entomologist ". . . just as the moon was rising, you could see the Indian Ocean as a narrow sparkling band, dividing the dark mass of the continent from the sky."

British rule too came and went. Tanzania gained independence and a new wave of African-led research brought with it excitement and groundbreaking studies—until funding ground to a halt. As Tanzania faced food crises and a war with Uganda, upkeep of the plantations halted. Research projects stalled, then slipped into paralysis. The scientists left, promising to return.

So, Amani waited. Three local technicians loyally tended to the mothballed institute in their absences. They swept the floors, cut the grass. For a time, the tiny post office maintained regular hours, even as damp climbed the whitewashed walls and rust poxed its metal sign. But mail dwindled to a trickle, and then to a halt. The library closed its doors, leaving a time capsule display of CURRENT PERIODICALS at the entrance. The technicians grew old and gray.

Of them, Martin Kimweri remains. A petite man with short cropped hair, he is now in his late fifties. Every day he lets himself into the dark space of the old laboratory to undertake his duties. Tattered curtains hang in strips over the windows. Laid out across a bench lie two dozen dead rats—roughly taxidermied, cotton wool poking through where their eyes should be—stiff with age. Behind them, chemists' glassware bears a dark film of grime. Formaldehyde-bound creatures decompose slowly in jars. Sample bottles, half-filled with pale fluids, stand ready for testing; hand-written labels date them to the 1990s. Water stains bloom across the plaster.

He passes through the corridor, feet tapping on the graying tiles, past a blackboard bearing chalk inscriptions: chemical recipes for cryoprotectant, buffer solutions. Kimweri lets himself through to the back room, where a plastic tank holds a colony of fourteen white mice, pink-skinned and red-eyed, on a bed of urine-soaked shavings eight inches deep. They have a thick animal stink. He feeds them, refills their bottle. He picks the mice up by their tails, drops them gently onto his lab coat. They cling tight with their little claws, let their eyes go wide.

In the past, these mice—or, mice like them—were infected with malaria before being treated with various experimental therapies. Curling posters on the wall offer diagrams as to their dissection. Kimweri has been feeding the Amani mouse colony since he joined the institute in 1985. There's not been much else to do since 1996.

It's possible, he thinks, that the scientists could still return. Maybe next year. Until they do, he'll be here. Keeping the colony alive.

Outside, thunder raises its voice. Rain runs down the warping panes. In this way, years pass.

Meanwhile, in the forest, a battle is unfolding in slow motion—one that was set in motion by the Germans when they transplanted the trees all those years before.

In their experimental plots, long left untended, incomers had at first bided their time. They put down roots and spread their arms, stretching out to their brethren as if for comfort. But, over time, there came a shifting in the ranks, a forming up. The new trees grew restless. They sent out emissaries, testing the ground beyond the edges of their plantations. No one was there to stop them.

The native vegetation was caught off guard. Before the Germans had arrived, with their barrowloads of plants from elsewhere, the vegetation of Amani had been more or less isolated for millennia. The Usambara Mountains, which form part of a wider chain of mountains stretching southwest from Kilimanjaro called the Eastern Arc, rise like dorsal fins from the depths. They rise to such heights above the arid, coastal plains as to exist in a climate all their own—one defined by its cool, damp mountain air where mist swirls and vapor rising off the ocean condenses as wet fog. Cloud islands, they are called.

Consider the Eastern Arc mountain chain as a forested archipelago, floating in midair: a Galápagos of the land, where, over millennia, species marooned upon their perches in the sky—separated from the rest of their kind by oceans of dry, infertile flats—split off along evolutionary paths all their own, like Darwin's finches.

A large proportion of the flora and fauna found in the Usambaras are thought to exist here and only here. The Amani flatwing, a delicate damselfly, is considered one of the world's most critically endangered species, and hovers along the banks of only a few rocky streams. The Usambara eagle-owl,[*] a striking, wide-eyed bird with barred feathers and a hawkish expression, was first spotted in 1908, and not recorded again until 1962. The Amani tree frog,[**] the sole species of its genus, was discovered hiding inside a wild banana in 1926 and never seen again.

All this is to say that these lonely hearts cling to toeholds in an ecosystem to which they are uniquely adapted; one that remained stable and undisturbed for millennia. This cloistered community has grown up together to create a forest so densely layered, so

---

[*]*Bubo vosseleri.*
[**]*Parhoplophryne usambarica.*

multitudinous, so variegated in texture, it is difficult for a visitor to parse it into its various elements. Trees burst from the bodies of their fallen ancestors. Vampiric strangling figs throw woody limbs around their victims' shoulders, bleed them dry. Epiphytes bed down in nooks of branches like birds building nests—collecting fallen leaves to make a compost of them. Fur coats of lichen drape luxuriously from dry-bark arms. A mass of heaving, verdant life.

Huge *Maranthes goetzeniana* trees rise from the mountain like Doric columns, dwarfing all else. Heavy with ferns, these silver-skinned sentinels began their lives in the age of the Vikings, and have been growing at a rate of inexorable slowness ever since. But these ancient organisms find themselves suddenly surrounded on all sides by strange new neighbors.

In the time it takes these giants to put out one new shoot, golden bamboo will come up in sheets, so densely packed it crowds out all other plants. It spreads sideways, under the soil, before bursting upward; cut it down and it will regenerate, disembodied limbs rooting where they fall. Sugar palms, docile at first, began to amass at the foot of the hill before charging up. Every year, they take a few yards of new territory. Their acid fruits poison the soil; nothing can grow under them; touch them and bleed.

Ornamental shrubs, planted in the gardens around the villas, spilled into their surroundings like pots boiling over, uncontained and uncontainable. *Lantana camara*, a pretty herb with floral puffs of pink and yellow, smelling of blackcurrant and beloved of butterflies, foamed outward, swallowing all that it touched. *Clidemia hirta*, with its elegant, fretted leaves and soapsuds smell, shape-shifted; taking the form of a vine, it clambered over the other plants, pulled them down.

Forests are never so peaceful as they appear. Before the new arrivals, native vegetation of Amani had fronts they had been fighting on for centuries: strategies developed and countered in a constant low-level struggle for light and nutrients. But these foreign invaders mounted a surprise campaign, caught them unaware. Known horrors had been turned loose upon one of the world's most isolated, most fragile, most biologically rich habitats in the world. Plants normally docile and well behaved in their home environments, finding themselves suddenly at an advantage, ran amok.

An alien invasion was taking place: slowly, silently. Outside, in the plantations, battles were being fought and won, while inside mildew climbed the walls, books rotted in the library, white mice were born and lived and died. And no one was paying attention.

Between downpours I walk the grounds, searching for the remains of the old imperial architecture. Alloyce Mkongewa, who grew up in the nearest village, accompanies me. He's a mild, good-humored guy in walking boots and a preppy polo shirt, who acts as a fixer for foreign biologists when they come to study Amani's escapees.

Together we wander the old spice garden by the entrance gate, where rows of mature trees emerge from a weedy tangle, Latin nomenclature spelled out on metal plates affixed to each trunk, like name tags at a mixer. The air is thick, heady; mulled like wine. A clove tree—*Syzygium aromaticum*—huge and bushy and untamed, offers handfuls of its bulbous green seeds. The nutmeg—*Myristica fragrans*—hangs heavy with baubles, each splitting apart to reveal a racy flash of scarlet mace inside. Black pepper—*Piper nigrum*—grows as a vine, swaddling the trunk of a rose apple—*Syzygium jambos*—its seeds hanging in bunches like grapes. Alloyce man-handles a cardamom—*Elettaria cardamomum*—as if it were an ani-mal, firmly parting its branches to reveal *Vanilla planifolia* wrapping succulent stems around its waist.

Farther back, in what feels like virgin forest, we stumble upon what appears to be a boulder with a metal pipe protruding; it's a water tank, smothered in mosses, and topped by a velveteen, flat-leafed fern which has made its home upon the lid. The tank was for the steam train that once moved along a track, right where I am standing. We walk out across an old concrete dam, now half-collapsed, where the Sigi River rushes through a gap in its base. Muscular roots of trees spill over its sides, as at the temples of Angkor Wat. Upriver, we find the rusted generator of an old hy-droelectric plant in a ruined hut whose window panes stand shat-tered in their frames. The infrastructure of a forgotten world.

Our progress is observed by a gathering of blue monkeys, sprawled on a raised platform of the invasive clidemia, which stretches like a sheet of leaves over a disturbed patch of forest, mask-ing the hunched forms of the trees below. "It's a huge weight burden," says Alloyce. "All the trees you see below will come down. Nothing else can get the light to grow. Things could get bad here, if it's allowed to spread."

At night I stay in a small, somewhat makeshift, rest house, not far from the old laboratories, set up for use by conservationists when the old Amani estate was cordoned off as a reserve in 1997. I find myself whiling the evenings away with the tropical biologists Pierre Binggeli and Charles Kilawe, who have done much of the studies of the invasive threats the Usambaras are facing. They form an unlikely double act: the charismatic Pierre tall and white, with mad professor hair and an international accent I can't place; Charles diminutive, hailing from the Chaga people of Kilimanjaro, a gentle presence, ready to laugh.

Pierre chides me for picking off the seeds sticking to my trousers and socks, and—unthinkingly—dropping them onto the tiles: spreading invasive species. (It's true: in the 1930s, the English botanist Edward Salisbury once raised three hundred plants from seeds found lurking in his trouser cuffs.) They are running a workshop for local smallholders—the Shambaa ethnic group, Alloyce's people, whose existence on the edge of the forest long predates imperial incursions—about the importance of guarding against aliens.

A rule of thumb: if a hundred species are planted in a new environment—a botanic garden, say—and then left to their own devices, maybe ten will become established. Of those ten, one will likely become a pest. But this is just a guide. Of roughly the 214 planted in Amani, a 2008 survey suggested that, in total, 38 exotic plant species had become naturalized, while 16 had escaped their designated plots and were running rampant through old-growth forest.

The fear is this: a surge of foreign invaders into an isolated ecosystem will disrupt food chains, alter soil chemistry, upset bacterial communities and mycorrhizal networks beneath the ground, outcompete slower-moving natives, carry in diseases. The current UK epidemic of ash dieback disease, for example—expected to kill 80 percent of all ash trees in the country—entered the country on an imported tree. Ash dieback is caused by a relatively common Asiatic fungus; its native bedfellows the Chinese and Manchurian ash—having coevolved with the fungus over thousands of years—are able to withstand its attacks.

Cautionary tales like these serve to underline the significance of long, entangled ancestral history. To paraphrase John Muir: when one tugs at a single thing in nature, he finds it hitched to the rest of the world. Tug hard enough, and we risk the whole beautiful tapestry coming undone.

The idea of coevolution—that two or more species developing in concert over long periods might come to slot together tightly—was first floated by Charles Darwin. In 1862, he received a consignment of Malagasy orchids, and found his interest piqued by the "astounding" *Angraecum sesquipedale.* Not for its beauty (although it is beautiful, with flowers "like stars formed of snow-white wax"), but for its intriguing "whip-like" nectary which hangs as a thin green spur around a foot long, the bottom few centimeters of which are filled with sweet nectar. He wrote to his friend J. D. Hooker in delight: "Good Heavens, what insect can suck it."

Darwin went on to predict a moth with a long proboscis that might unfurl to the same outrageous length, each having spurred on the other to increasingly specialized adaptations, until no others would have them: an evolutionary love story playing out over millennia, a folie à deux. A moth matching this description was finally confirmed to pollinate the orchid in 1993, a hundred and thirty years after Darwin's prediction. There was the lock, and here was the key. It slipped right into place.

But what Darwin saw as evidence for his mechanistic theory of evolution, others felt was convincing evidence of intelligent design. Indeed, to consider an ecosystem in all its beautiful and mind-bending complexity—especially those long isolated, like the inland islands of the Eastern Arc—is an exercise of an almost transcendental nature. One feels keenly the sense the interconnectedness of all living things; a sense of everything being in its right place. This is biology as teleology: rainforest as a pocketwatch fallen open on the path—but instead of the hand of God crafting its workings, coevolution is the force by which the cogs of an ecosystem shape themselves to one another's tines. Or so it seems to me. (Wrote Darwin, on God: "Let each man hope & believe what he can.")

Seen through this light, the arrival of new and invasive species comes as a spanner in the works: too many outsiders blundering in could spell chaos, as carefully calibrated symbioses and evenly weighted relationships are thrown off balance. Such is the sentiment underlying the field of invasion biology, the study of human-led introductions of organisms to new places outside their natural habitat, as has been the case many times over in Amani. As a field, it traces its origins to the Eeyoreish Oxford don Charles Elton, who published his groundbreaking *The Ecology of Invasions by Animals and Plants* in 1958, and has seen an almost exponential growth in publications since the turn of the twenty-first century.

Invasion biologists have made study of some truly horrifying cases of what can happen when invasive species are allowed to rampage out of control. Islands, shielded from the evolutionary pressures of the outside world, are particularly vulnerable. On islands with few or no predators, residents let their guards down, leading to strange varieties of life—"flightless birds, thornless raspberries and scent-free mints"—which prove hapless in the face of predation or more sharply honed competition.

On Gough Island, for example, a remote and treeless outpost in the South Atlantic, the harmless house mouse has caused a bloodbath. Inadvertently introduced by sailors in the nineteenth century, the mice have grown super-sized in the absence of predators and developed a tasted for seabird chicks—including those of the Tristan albatross, three hundred times their size. Two million chicks are lost to the mice every year, and without intervention the albatross will likely go extinct.

On Guam, the brown tree snake threatens to bring down an entire ecosystem. The snake, introduced by accident in the 1940s, eats forest birds—ten out of twelve native species have been lost, the remaining two functionally extinct—and without birds to spread their seed, the trees too are thinning out. For the same reason, 61 percent of all extinct, and 37 percent of all critically endangered, species are from islands.

To an outsider, however, the language of invasion biology dovetails disconcertingly with the language of empire, with its uncomfortable, even xenophobic, overtones. Plants indigenous to a place are described as "natives"; those recently arrived are "alien" or "exotic," sometimes "colonizers," or "invaders." The tenor of the conversation does have a certain concord with its subject matter— so many of these invasions, after all, being attributable to and enabled by colonialism. In particular, imperial botanical gardens like Amani have historically been some of the worst offenders in enabling the spread and establishment of invasive species.

The earliest mention of *Rhododendron ponticum* in the British Isles—that glossy leaved, extravagantly floral ornamental, which has become "the most expensive alien plant conservation problem in Britain and Ireland"—can be found in a 1768 list of species then cultivated at Kew. The triffid-like giant hogweed, which grows to twenty-two feet tall and has become notorious in the tabloid press for its noxious, burning sap, was also first recorded at Kew. It counted among the seed collection there in 1817; by 1828 it was growing wild in Cambridgeshire. (One contemporary newspaper ran a cartoon of the plant climbing over the walls and jumping on a number 27 bus.)

A 2011 study showed that, of the thirty-four plants listed by the IUCN as among the hundred worst invasive species worldwide, more than half are known to have escaped from botanical gardens.

The dumping of surplus water hyacinth specimens from Java's famous Bogor Botanical Garden into the Ciliwung river, for example, saw the notorious pest storm across the island, clogging waterways and rice paddies. Lake Rawa Pening, five hundred miles away to the east, is now smothered by a weed mat covering 70 percent of the lake's surface. The situation bears more than a passing resemblance to the literal alien invasion in H. G. Wells's *War of the Worlds*, in which a noxious alien plant spreads quickly from the Martian crash site to take over all London, rendering the Thames a "bubbly mass of red weed."

The IUCN "least wanted" list—the outlaws of the natural world—are the textbook cases, the ones that, when invoked among conservationists, bring a chill into the room. They are the proof of the law of unintended consequences, terrible warnings of the dangers of interfering with the natural order of things. At least three of these reviled species have escaped from a single source: Amani, Tanzania.

In *non*-abandoned sites, great effort is often invested in the control of such weeds and pests. Numbers are "managed"—that is, eradicated, killed off wholesale by way of weedkiller or poison pellets or sharpshooters. The question of "population control," of culling, of extermination, is one of the biggest ethical quandaries at the heart of contemporary conservation. Charities and researchers often advocate an end to certain forms of wildlife in order to favor certain other forms of wildlife—killing gray squirrels in favor of their red cousins, to give a British example.

New Zealand, an island nation with many strange and vulnerable endemic species to protect, perhaps takes this further than any-where else. It spends $42 million to $56 million on pesticides, bait, traps, and poison-dropping helicopters each year; in 2016, the country announced plans to eliminate all rats, stoats, and possums by 2050 in order to protect native birds like the kakapo and the kiwi in one mega-eradication—involving the deaths of many millions of small predators.

This is conservation as interventionism; killing by the strong to save the weak. But, as with foreign policy, such strategies are often contentious. Their moral rectitude too is a matter of opinion.

In Amani, it may already be too late to halt the progress of the aliens. From the red-mud road, everywhere I look, I see a coiling mass of *Clidemia hirta* or *Lantana camara*. Deep in the rainforest, Spanish cedars and Mexican rubber trees proliferate among their fellow trees. Once noticed, one can't unnotice them. They lurk in the background of every scene. It begins to feel unsettling.

Such is the dilemma of conservation bodies seeking to preserve the unique biodiversity of the East Usambaras. How can we hope to remove that which is already ubiquitous? How can you root out one or two types of tree, without disturbing all else in the forest?

It's very difficult, I know. Back home, my friend Louise directed a project in the Highlands which sought to remove the *Rhodo-dendron ponticum*—descendants of those escaped from Victorian gardens—from where it had been spreading beneath old-growth Scots pines and crowding out the native species. It was, effectively, all-out war. For the best part of a year, the team ripped out rhododendrons, chainsawing them into sections, then feeding the cuttings into a mulcher, which spat them out as chippings. In other places, they built pyres and burned the corpses. Afterward, they injected the stumps with herbicide, to prevent resprouting.

It was a violent process, and the end state looked like a battlefield. Disembodied limbs lay scattered over churned-up mud. The ground was wet and raw, opened to the sunlight for the first time in generations. The old pines shocked and naked from the waist down. But there was a sense that the battle, though grueling, had been won.

In spring, I returned to inspect for progress. Tiny seedling pines were bristling, tufts of grass were skinning over the raw flesh of the land. But my eyes were caught by these smooth, strong shoots of acid green pushing through from under the chippings with blind, snake-like heads; pale fronds of tiny, unformed leaves budding at the tip. To truly rid the estate of rhododendrons, it was clear they would have to kill them all over again. And again. And again. "It will never be over," said Louise.

Louise is keeping her word. But it's one thing to firefight forever, in a small corner of an estate long stewarded by man. Another entirely to transpose the same all-guns-blazing tactics into the cloud islands—refugia in the sky where strange creatures take cover in dark crannies. In Amani, where the alien fugitives have run free for so long, one fears the horse has bolted. The abandoned plots are merely the original encampment of forces long gone native.

On my last afternoon in Amani, I take another walk with Alloyce. We pass along a track behind the laboratories, near the old boating pond. Now long drained, it is a flat footprint in the forest, dense with native ferns and noisy with the voices of a hundred frogs. The trees are skewbalded here with a flat lichen that grows in uneven patches of pale green and khaki, some the size of fingerprints, others handprints, which—combined with the sun dapple through the canopy—create the illusory sensation of both depthlessness and infinite depth; my eyes swim in the mid-zone, searching for an anchor. A few older specimens wear heavy lianas like pythons draped around their shoulders. Ferns and creepers wedge themselves in every crevice, throwing down ropes.

Not much farther we come across a gap in the forest cover, where the original trees had been logged a decade or two before. The ground is not bare, exactly, but dressed with a thin and leafy gauze, like camouflage netting, pierced by thin, young-looking trees all of around the same stature. Alloyce identifies them: *Maesopsis eminii*. Self-seeded. Non-native.

Back in the 1980s and '90s, a lot of the anxiety about invasive species in Amani centered around this tree. Scientists, including Pierre, became concerned by how quickly maesopsis was spreading through the forest, far beyond where it had originally been planted, outcompeting native species. It frightened the scientists: it was the first example of a tree invading a continental location. Their worst-case scenario predicted maesopsis might come to invade half of all the natural forest in the region. If so, soil fauna, water run-off, and all sorts of basic functioning could be sent haywire.

But what could be done? The abandoned test plots had been declared part of a nature reserve, and were protected from counterattack with the herbicide I saw used against rhododendron. Now they were effectively abandoned to their fate for a second time. By 2011, maesopsis had come to comprise a third of large trees in secondary forest in the region, and 6 percent even in "more pristine" forest. "*M. eminii* may now be too widespread for eradication or even local control to be feasible," concluded researchers.

However—though widespread—the maesopsis uprising had failed to materialize in quite the way that had been anticipated. Once the first wave of maesopsis matured, it began to become clear that its reproduction rates had slowed. In the increasingly well-shaded forest floor at their feet, their own seedlings were losing ground to the natives, who now were spontaneously regenerating and reclaiming their land. In providing shade—it could be argued—it has served a useful role. It too feeds local fauna: hornbills, who feed on the maesopsis seeds, have boomed.

And it's not the only alien carving out a role in their new home. Those frightening black-armored millipedes I saw earlier, imported from New Zealand, have made a seamless transition to the

Usambaras, and serve a role not unlike that of a dung beetle. Even the detested lantana and clidemia have their fans: butterflies and the endangered Amani sunbird, with its iridescent green shoulders, gorge on their nectar.

That non-native species should be able to settle in—and even come to serve some helpful roles—does throw a little cold water over the idea of the ecosystem as the intricately wrought, carefully balanced product of millennia of coevolution, each species carving its way into the genes of the others' as water sculpts rocks. Indeed, it is the success of aliens like maesopsis, all over the world, that has fed into a bold new perspective, one that seeks to reevaluate environments "tainted" by the proliferation of non-native species.

Such ecosystems, suggests this new school of thought, are "novel ecosystems"—created by man, but self-sustaining. And it is time to accept the fact that they have changed forever, cannot be returned to a previous "unspoiled" state—indeed, that the notion of their being spoiled in the first place may be incorrect—and appreciate them, for what they are now. It is, write its proponents, "the new ecological world order." They quote Goethe: "[Nature] is ever shaping new forms: what is, has never yet been; what has been, comes not again."

Some of the important early work on novel ecosystems evolved out of the work by Ariel Lugo in Puerto Rico during the 2000s, who studied regrowth on abandoned sugar, tobacco, and coffee plantations (another landscape with roots in the colonial era). As in the former USSR, the twentieth century has seen large-scale regeneration of forest in Puerto Rico: forest cover rose tenfold to

60 percent. But these new forests were a mongrel assortment of various introduced species: mango, grapefruit, avocados, and the blazing coral flowers of the African tulip tree.

The response to the new growth was lukewarm; some local conservationists reportedly proposed chopping the whole lot down and starting again, with plantations of favored native species. But over time, it became clear that something wonderful was happening. The invaders were merely paving the way for a more generalized environmental recovery.

As the hornbill feasts upon the fruits of the maesopsis in the old Amani hill station, the tulip tree forests that grew up in the abandoned Puerto Rico plantations became home to native species, including a small endemic frog called the coquí, which has a piercing cry. Its numbers had fallen into decline, but its whistling call now again pierces the air of the mongrel woods. The new forests, reports Lugo, support greater biodiversity than the original forests; partly as a function of the huge influx of non-native species (like parakeets and the blue-and-gold macaw from Paraguay) who rub shoulders with native finches in their new shared home.

But Pierre, my tropical biologist friend, is reluctant to kowtow to the new world order. "Most such systems are species poor," he warns. And as for rare species, their situation is already so parlous that to welcome major changes to the ecosystem would be reckless.

Generally, critics feel that by embracing novel ecosystems, we are abandoning hope of undoing the damage done by humans, or offering a free pass to companies or governments that damaged them in the first place. Still, with around a third of all ice-free land now thought to be covered by novel ecosystems, it grows increasingly important to wrestle with what these mongrel, immigrant communities mean for the future of our planet as a whole. And it is in abandoned places, where human-impacted land is *not* being managed—where non-native species and native species alike are left to their own devices, without heavy-handed but well-meaning intervention—that we might begin to view alien invasions over a longer period of time, and perhaps come to to appreciate that, in time, an ecosystem might start to adapt to its new citizens and find a new sense of balance.

Invasions of exotic species often display a "boom" period of exponential growth, followed by steep and sudden decline. The aliens may overexploit their new habitat and run out of food; once-hapless prey adapts to their new predators; or pathogens in the new environment might get the better of them. On Santa Cruz in the Galápagos, for example, the cinchona (quinine tree) escaped from its plantations in the 1940s and came to dominate forty-two square miles of the previously treeless island. It shaded out native plants and altered nutrient cycling in the soil. It was an ecological disaster. Then, around a decade ago, the cinchona began to sicken from an unknown cause. During the subsequent years, the trees have been dying back almost as quickly as they arrived, leaving the island pronged with deadwood—and opening it to "substantial regeneration" of the native species beneath.

Such was the fate of H. G. Wells's red weed too. "In the end the red weed succumbed almost as quickly as it had spread," wrote Wells. A "cankering disease" caused by terrestrial bacteria, of the sort regularly fought and resisted by every species on our planet. But the red weed had no defense. It "rotted like a thing already dead": "The fronds became bleached, and then shriveled and brittle. They broke off at the least touch, and the waters that had stimulated their early growth carried their last vestiges out to sea."

There are signs that this too may happen in Amani. Sometime in the last ten or fifteen years, a strange growth began to appear on the smooth skin of the maesopsis trees. It was a bracket fungus—disk-like shelves protruding from the bark, rotting their hosts from the inside out. Initially few and far between, the fungi grew and spread, and is now killing maesopsis where they stand. It appears to be a native, says Pierre, although its true origins and the key to its sudden spread are not known. To some, it might look like revenge.

The walk back to the laboratories is long and steeply uphill. Alloyce and I fall silent, passing through the various nations of plants. It feels like a zoo, all the gates left open.

The earth is sandy again here, almost ochre, but elsewhere it fleshes out and bears a rich, almost crimson tone. The grass itself seems to flinch from my step. I pause and realize that my eyes are not deceiving me; it is *Mimosa pudica*, "the sensitive plant" known for its cowering demeanor. I touch one with my fingertip and its tiny leaves cringe. I walk on, with an apologetic gait, and see the shockwaves travel from the plants closest to my feet outward—villagers fleeing indoors from a blundering giant—pulling closed their shutters and battening down doors. As well they might. In many countries, including Tanzania, *Mimosa pudica* is considered an invasive weed and dug from the earth.

In so many places, we are so busy playing at being stewards of the Earth, deciding who gets to live and who gets to die. Once we have left our mark on an ecosystem, we show no hesitation in throwing open the hood again later to fiddle with its workings. We run the Earth as if it were one giant botanical garden to tend; passing judgment on species, playing God.

I feel nervous to think of it. It seems an imperial line of thought, one centered upon "improving" and "civilizing." And my time studying colonialism in Australia has made me cautious. I know well the ways that interventionists with the best of intentions can cause as much harm or more than those with poor ones. At what point must we learn to let go, and watch the repercussions of our past actions spin out into the void, and give the Earth its head to respond and adapt in the ways only it knows how?

In the afternoon, with time to kill, Alloyce takes me on the back of his motorbike along a dirt track out of the reserve. We pass through a Shambaa village and then another, where bright kangas in every possible color are stretched out over bushes to dry. We weave uphill to the shell of a derelict coffee factory long closed—more foreign interests failed and forgotten. The former director's house stands on a precipice. It's a shell, missing windows and doors, scrawled in graffiti. I stand with my back to it, feeling again the vertigo of the cloud islander, looking out over dusty plains to the coast. A bathymetrical view, staring down into the depths of an invisible sea. There's not much to say.

After a while, we turn and jolt back down the track, my hands on his waist. As we go, the weather breaks: great sheets of water, a cloudburst, an avalanche of rain. When we reach the first village we run to shelter on a concrete veranda. Three villagers are waiting there already. They nod hello, but no one speaks. We stand there in silence, watching the rain hammering down, forming up in lakes and rivers and floods where before there was only dust.

I am soaked to the skin, my clothes clinging to my body. The rain dies to a drizzle, and the cloud begins to thin and let through the light and we see it, water hanging in the air, sunlight cutting in shafts between the trees, everything glistening. We stay there some time. Watching, listening, waiting for the rain to end.

10

THE TRIP TO
ROSE COTTAGE

Swona, Scotland

W hen he drops me on the island, Hamish the boatman has a last piece of advice: "Stay in the house at night," he says, "and lock the door behind you."

"Oh?" I say, taken aback.

"Don't camp outdoors," he repeats, "or the cows will trample you. Make sure you sleep in the house. See you tomorrow." Then he's gone, and I'm left alone on my desert island. Just me, and the birds, and these trampling cattle. I turn to face it: green and tumbledown and wind-battered, and feel for the first time a shiver of unease.

This is Swona, a small island off the tip of the northernmost point of mainland Scotland. Though always on the peripheries, it is a place with a long history of human occupation. The chambered cairn attests to the presence of Neolithic farmers in 3500 BC or even before. Perhaps four thousand years later Celtic missionaries came ashore in their coracles; the Norse arrived sometime in the ninth century, and their descendants were still there a millennium later. People washed up here, and remained.

Different names, different tongues, but more of the same, if you get down to the nuts and bolts of living: they tended their livestock, they grew barley and oats in a patch of fertile ground, they grew rhubarb and potatoes in the shelter of low stone walls, they built boats. They fished for coalfish and dogfish and dried them in the salty air. They kept cows: domestic cattle, of the sort that too can trace a domestic lineage back to Neolithic times.

By the eighteenth century, there were nine families on the island, planting the same small strips of land, building new houses from the stones of the old. Names on the census flowed continuously from decade to decade, generation to generation: Halcrow, Gunn, Allan, Norquoy, Rosie. Island life flowed on, century to century, much the same as before. Until, suddenly, it didn't.

In the 1920s, the fish market collapsed, and many in the community lost their main source of income. Off the island, the world was changing fast. Rather than stay there, stranded on this rocky outcrop, bound on all sides by breakwaters and whirlpools, eddies and races, many chose to leave.

Some went just across the water to South Ronaldsay, the most southerly of the Orkney Islands. Others moved to mainland Scotland, visible across the water to the south if you were to stand on

Warbister Hill, the island's highest point. Others decided to emi-
grate, and try their luck at a new life entirely: there was a whole
world out there, one that they had seen only a corner of. By 1927,
the Rosies were the only ones left.

How would you live on your own private island? This is how
the Rosies did it:

They kept chickens, they kept cattle, they kept house. Their
five children ran wild, scrambling in the rocky gullies and pad-
dling in the shallows (although they didn't climb trees, because
there are no trees). They weeded and darned and mended nets.
They scavenged on the shoreline for items washed up—revelations
from the outside world. They read everything they could get their
hands on. They wrote letters, and received them too: handwritten
notes addressed simply "Swona," or sometimes to their house,
which was named for them: Rose Cottage. They played instru-
ments, and formed for a while an island orchestra with fiddle,
pipes, squeezebox, and a makeshift set of drums made of oil cans.

Their father built boats, manned the tiny lighthouse, and in
1935, after a cargo ship ran aground off the western shore, salvaged
enough from the wreckage to install electricity in their house—
powered by windmill and a diesel generator. After that they lis-
tened to the radio: news, plays, tunes, the shipping forecast.

They grew up. They grew old. Mr. and Mrs. Rosie grew older,

then infirm, and then they died. Two of the daughters got married and left. And then there were three: the twin brothers Arthur and James, and their sister Violet.

In 1957, journalist Comer Clarke paid a visit to the island and in his subsequent article made much of "the Silent Woman of Swona": Violet, he claimed, had not spoken to anyone outside her own family for twenty years. Not much call for it, I suppose. The family protested: she did speak, they said—albeit in a whisper, and to those she knew well. In any case, she lived happily enough, in spite of the unwelcome attention, on this small island alone with her brothers and their animals.

But over time, the siblings too grew old and frail. Arthur died in 1974, and then there were two. By that time, James's health was ailing and, soon after, they sent a distress call to family on South Ronaldsay. James and Violet packed their belongings, taking only what could be carried, bundled together in sheets and tied up with ropes, down to the Haven where a boat had been sent for them.

Last of all, almost as an afterthought, they turned to the barn and opened the gate, letting the cattle loose, to fend for themselves until their return.

The sun sets and rises, sets and rises, chasing shadows across the floor, across the table, across the wall. Rain lashes the windows to one side, salt spray the other. Days pass. Weeks pass. Months. Years.

Outside, torrents rush past the rocky shore. The lighthouse to the south and the beacon to the north pulse out automated advertisements to their presence. Constellations whirl overhead. The moon waxes, wanes, waxes, wanes, waxes. The cattle live, give birth, and die.

Inside Rose Cottage, dust settles unseen. At first it forms a thin veneer, but then a thicker, felt-like scrim, pulled up and over everything. It forms over tea towels hung to dry over a stove long ago gone out. It settles on the coal in the scuttle and the kitchen table sitting ready for a meal, with a jar of marmalade, tinned milk powder, and a box of biscuits at its center. It settles on the papers stacked in piles on the sideboard, and on the sewing machine packed neatly in its box, on the ham radio by the window and the stopped clock on the mantelpiece reading ten minutes past three.

Later, as the damp sets in, the air grows thick with decay. Tins pox and swell in the cupboard where they were stockpiled. Glassware takes on an opacity—that hazy, milky quality of age—and the mirror a patina of gray-green that creeps in from the edges, clouding the reflection. Salt in the shaker solidifies into a single, molded block. Upstairs the beds are still made, ready to be slept in, the sheets pulled neatly up and tucked in tight.

Just over a decade after the Rosies' departure, the photographer John S. Findlay came to document the island, and noted that the sense of human presence was still so strong as to prompt him to knock upon every door before he entered. The feeling that the owner was only in the next room, or shortly to return and catch him, was intense. At that time, the house still resided in a realm of mere absence—as if someone had slipped out for a walk—although a few artifacts suggested the start of the transmutation of this absence into something altogether more profound.

By the time I enter the house, more than three decades after that, the metamorphosis has advanced. Now, it has clearly been abandoned for some time. There are still traces of how things were left—the wipe-clean tablecloth left in place, though its laminae are separating, its skirts shredding onto the floor; the soft furnishings rotting away to bare wooden frames; paperwork stacked, but soaked and softening to pulp—but the next phase, ruination, is now surely close at hand.

A newspaper rests on the table—wet and flimsy and folded in half. Its uppermost pages are too perished to be readable; when I tentatively lift its corner with one finger I find the cover hiding inside: a *Press and Journal*, bearing news of a change in government: Ted Heath out, Harold Wilson in.

I carry in my phone a photo of this room, taken by Findlay in 1985, and I bring it up now to chart the advance of its decay. Items are missing: that stopped clock, for one thing, with an art print propped up in its place. The doors to the cabinet have been flung open, the paperwork riffled through and stuffed back. A kettle—aged, rusted—has appeared on the stove top. Though many years have passed, the suggestion of unseen presences moving in the space between the photo and the scene before me is uncanny.

A tide of mud has swept in under the door and across the floor. It has settled in a layer of around an inch thick, soft and wet still, perhaps from the storm that's just passed over—the one I had to wait out in a tent on South Ronaldsay for two days, canvas flapping through the night, the whole setup threatening to take off. In the mud, a broom lies half-submerged and dropping its bristles. And next to that, a set of footprints that are not mine.

These too catch me unaware. I feel a sudden chill. I freeze where I stand and peer through a loose lattice of floorboards into the bedroom above, then down again to the footprints. It's not clear how long they have been there. It's sheltered inside the cottage, with the door shut. They could have been here for years, like footprints on the moon. Although—they look fresh.

I speak out loud without meaning to: Hello?

No answer. The house presses in silently against me.

It's obvious which house Hamish meant me to sleep in. Only one stands firm, one of the Rosies' neighbors. It's a broad, stone-built building, more solid than handsome, its exterior intact—or, mostly so. Two roof tiles are missing to the rear: the kiss of death. Without repair, ruination is now inevitable. Until then, this is my best hope of shelter.

When I try the front door, it seems to have been locked from the inside. I throw my weight against it. Nothing. I drop my pack to the ground, and think about leaving it there—leaving the question of where to sleep for another time. But with a curtain of rain sweeping in from the sea, I circle the building, briefly consider clambering through a small window, before finally managing to jiggle loose the door to a lean-to at the rear.

At first it seems it's only outside storage, with no way into the main house, but an internal door lies hidden behind a huge sheet of fiberboard. The whole place seems designed like an obstacle course, and as I step in I am half-braced for a booby trap. I find myself in the kitchen: an old gas cooker stands in the corner, its white front splattered with what looks like ink; an assortment of empty jugs and other vessels are lined up along the mantelpiece. The air smells stale, dead. I rest my backpack on a chair, sitting it upright as if it were a friend.

Outside, the rain has already passed by and the sky is clearing, the wind blustery and fresh, and I pause to watch the shadows of the clouds sliding fast and frictionless over the land through a grimy window. Here too the walls are stained the green of fresh clover, painted on in uneven strokes: algal-dark in patches, others still the pale powder-white plaster. The air is thick with dust, motes of which rise and fall in asynchrony through shafts of light. I think: *the skin cells of its previous occupants*, and find I can't unthink it.

The house has a traditional design: each gable has its own chimney and fireplace, with a narrow hall between. I find the front door I'd tried so hard to dislodge from outside has been barred by way of a heavy post roped to the handle, which strains against the doorway, holding it fast. It's an eerie scene—so devoid of people and yet so thoroughly fortified—like the calm before a battle, or after one—when it's too late to save anyone.

The fortifications aren't for me. They're for the cattle—and the birds, and the seals. Animals find their way into buildings like this, a door closes behind them, and they are trapped. They starve that way, or dash themselves to death against the walls. The greater part of Rose Cottage is long boarded up, after—years ago—a cow perished in what was once the parlor.

A narrow wooden staircase leads upward, and as I take it I remember another of Hamish's warnings: the last time he came out to look at the island, a few months previously, a tent had appeared in a bedroom of this house. An orange tent. He'd called out to check if anyone was inside the tent, but no one had answered.

"What?" I'd said, startled. "Why?"

He didn't know. He hadn't taken anyone else out here. I could sleep in it, he suggested.

The idea of it—of approaching an unknown tent, inside a house—of unzipping it, and peering inside to check for occupants, filled me with horror. And the thought of lying down in the dark inside it, of listening through the canvas into the house at night beyond . . . No.

Anyway, the point is: when I get upstairs the tent is gone.

The cattle are still here, though. I haven't seen them yet, but the evidence is everywhere. Not far from the house I find a flat expanse of mud which must sometimes form the floor of a shallow pond but is currently bare and cracked and scattered with sea litter—scraps of rope, plastic bottles, lost buoys—and clods of dry manure. Its edges are trampled, churned-up turf, dried to a solid moonscape.

I stagger over it, past the rusted ruin of what was once a tractor (half-submerged by earth, its wheels fallen away) and an old winch, to which the remains of a wooden boat is still attached, its bleach-white prow pulled safely from the water, but disintegrating completely at the stern, where countless beasts have rubbed their hot bodies.

A tumbledown wall leads me through a verdant patch of flag irises, in the process of unfurling their golden pennants, and marsh buttercups, then through a frizz of sea mayweed and wild chamomile. I move awkwardly, dogged by the sense of being watched.

In a way I am. Since my arrival, two oystercatchers have had me under close surveillance, flying up on my approach and retreating backward in a bounding motion, reluctant to take their eyes off me. They bounce shrill notes of complaint between them at fraction-of-a-second intervals, making it impossible to relax or even to hold a thought in my head for more a few moments. I am growing increasingly flustered—the effect is not unlike that of having set off a car alarm in a confined space. I hush them, call out to them—I'm here, I'm here, I mean you no harm—but they pay this no heed and continue their alerts.

Life shrinks away from my approach—a wren flees from a gap in the wall as I pass, revealing its nesting place, and as I approach a clutch of crumbling buildings, each sitting along a continuum from ruin to wreck, a crowd of starlings burst from their walls, heckling with throaty voices as they take the air. They move as a shoal, shimmering as they flip and switch as one and come in to land on the ribs of a nearby roof.

The main cottage has long ago lost its windows and doors, but its exoskeleton remains firm, haloed by a fluoro-orange lichen radiating from what roof tiles remain. To enter, I must step high onto what seems a raised floor—one that gives under my weight with a sickening lurch. As my eyes adjust, I see the cause: the room is flooded with slurry—thigh-deep and set in places to a smooth, level surface like poured concrete. Under a gap in the roof where the rain gets in, it has liquefied and dried many times over, forming a warped and dusty crust pronged through with fallen beams. The manure has risen up the sides of the interior walls, filling the fireplace and truncating doorways, and is in the process of spilling, in slow motion, through the front door onto the ground outside.

It does smell, but not as badly as you'd expect. Like soil, or wet vegetation. The cattle must use this space as a shelter, a sort of makeshift barn. The interior is lit by a diffuse green light. Sunlight filters softly through the rafters, striping and patching the floor; whitewashed walls are long grown over with algae the color of apples and emeralds.

It's quiet in here, with the bluster of wind muffled and the continual peal of the oystercatchers fallen away. The only noise is the wash of the waves upon the stony strand beyond, dampened to its barest elements: wash and withdraw, wash and withdraw, shuddering through the ground as a heartbeat. It's peaceful, church-like: a place of the cows. I can feel their presence here keenly—warm bodies pressed together in winter, hot sweet breath condensing in mist over their heads.

I have to duck to exit, and that's when I see them: silhouetted on the brow of the hill, the familiar unmistakable gait of domestic cattle. I try to count them as they move down the slope toward me but they pass in front of and behind one another, making it difficult. Fifteen, I think, including two calves.

I step from the doorway to get a better look at them, and the animals at the front see me, stop short. Those at the back stream forward and the herd bunches together. As awareness of the intruder spreads, they move to encircle the calves between them. They are chestnut, black, white-gray. I recognize them.

When I was a child, my family moved into a house surrounded on all three sides by a field, which would be populated alternately by sheep or by cattle. I liked the sheep well enough, particularly those that had been bottle-fed as lambs and never lost their affinity for humans. But the cattle were my favorite. Friendly and docile, but wary enough not to crowd you in the field as horses sometimes will. Once, during one of those long endless summers off school, I devoted weeks to familiarizing myself with the herd of heifers on the other side of the fence: chatting to them, telling them off, hand-feeding them grass, touching and stroking their faces.

They were sweet natured, jovial, just as curious of me as I was of them, and we explored the boundaries—each flinching away if the other overstepped an invisible line, but then returning again—building up a kind of trust. I came to recognize them individually, that fermented honey smell of them.

It's this that comes to mind when I see the cattle approach. I feel a surge of warmth and familiarity that fades to dissonance as they freeze in place. A white cow at the front steps forward and stares, almost defiantly, in my direction.

"Hi," I say. Animal to animal.

I don't say it loud enough to travel, but they turn and leave me anyway.

These cattle are the descendants of those let loose that night in 1974. Forty-six years have passed and, as with Rose Cottage, since then the nature of the cattle has been changing—from unobserved domesticate to something else, something feral. Something increasingly *wild*.

Old photos show the Rosies' draft animals huge and fleshy, bracing willingly into the collar, pulling their master on a ride-on harvester. Those spared physical labor were pets of a sort, handled daily, kept for their meat and their milk. Each had a name, and a place where it stood in the double-gabled barn. Familiars, then. Docile, biddable creatures of habit and careful training. What must they have thought when they were first turned loose and left to their own devices? For how long did they wait by the barn for their milking and their feed?

At first, relatives of the Rosies—their brother-in-law Sandy, then his sons after him—assumed stewardship of the island, expecting to work the livestock alongside their own. In the early years they made a few attempts to castrate the bullocks, and later to sell them at market, herding them into a rocky inlet or ruined house, then manhandling sedated animals onto a boat.

But this was a process that required at least six men—three to control the boat, then three to every animal, with halter, neck rope, and stick. Within a couple of years it became clear it was too much work for too little reward. Worse, nearly all the animals removed in this way died, due to stress or diseases they'd never built a resistance to. Accustomed to looking after themselves on the empty island, they were no longer gentle giants, but heavyweight brutes who shrugged off attempts to be contained.

The Swona cattle today are not so much aggressive as highly defensive, and more than capable of charging should a visitor—like me—be so daft as to get too close. This would seem to be a feature of cattle gone wild; in southern California, another rare feral herd—this one numbering around a hundred and fifty—has recently become notorious for terrorizing hikers on the popular Pacific Crest Trail. Though these cattle have peacefully grazed the region for a century or longer, a spike in visitors to the area has seen attempted gorings and calls for the herd to be eradicated.

On Swona, attempts to make even an annual roundup were abandoned and the animals were left to live as they wished—roaming the island, finding shelter among the increasingly broken-down ruins. When the grass grew pale and sickly in the winter months, they scratched in the rocky inlets for seaweed.

Though there was always the expectation they might die out over a harsh winter, it never happened. In fact, they thrived. From the original complement of eight cows and one bull, numbers expanded to thirty-three in ten years. And though their condition dropped off through those long bleak northern winters, sometimes resulting in deaths, the vet—who still visits the island once a year, to inspect them from a distance—found them generally in good health.

Quickly the cattle became a scientific curiosity. At the very least, their behavior—entirely unmanaged—was of great interest. As one of an extremely small number of truly feral herds in the world, it wasn't clear how they would act. All males were left un-castrated, and allowed to live to maturity. Breeding took place freely. Such things would never take place on a farm. Without the strictures of husbandry, they were forced to improvise. Now, their social organization is almost completely unrecognizable; much of what we consider to be "cow-like" behavior, it transpires, is not necessarily their true nature.

This has been an experimental, non-linear process, a reversion to primal instinct. Among the females, power struggles arose: a dominant cow would come to the fore, making decisions on where to graze and where to shelter. And as more bullocks were born and came into sexual maturity, they were brought into competition over the females and the right to breed. From their number, only one would be crowned king of the bulls, and it was he who would produce from among his offspring an heir. As with deer or wild horses, other males that tried and failed to challenge their ruler would be driven from the herd and exiled to the outer reaches of their kingdom. And, as the king aged and his powers waned, he too would one day be deposed.

When the animal scientist Professor Stephen Hall visited them in 1985, he found the cattle living in a single herd that roamed all the island, except for the northern headland, "which was the preserve of [an old black bull] who seemed to have been 'banished' there." This bull, likely a fallen king, was now living out his last days in obscurity.

Numbers crashed again during the 1990s, at which point the sex distribution got badly out of whack, and for a while it seemed like the survival of the herd was at risk. By 2004, there were ten bulls but only four cows, ensuring extreme levels of competition between the males. Visitors found the island scarred with patches of disturbed earth, where bulls had been violently pawing and beating at the ground in displays of aggression, and the island shook with their battle cries. A bull's roar is low and guttural, a heaving moan of anger and frustration that begins deep in its throat, and rises to a great reverberating crescendo, before falling away. He might turn his head to the sky, bringing a change in timbre, and break into a breathy, creaking bray, punctuated with snorts and bellows as he kicks up dirt and slams his hooves in a demonstration of his great weight and strength and rage.

During this period of civil strife, as many as four bulls might be banished at any one time, and would roam together or apart on the bleak headland where the beacon beats its rhythm through the night, and where the terns build their nests on the churned-up grasses. Alone with their testosterone, their frustrated ambitions.

They were exiled, but not forgotten. In 2013, John Findlay returned to the island and witnessed an event of great significance among the cattle: the death of the banished king.

> *I became aware that the old black bull was lying on his side on the ground some distance from the herd. He looked dead but the odd twitch of his tail indicated that there was still some semblance of life . . . About an hour later when passing Rose Cottage we became aware that a number of the cattle, being led by the young black bull, had left the main herd and were making for the old bull who was obviously in a distressed state. The group certainly gave the impression of being genuinely concerned and were nudging and making physical contact, providing some form of comfort to him in what was a dire situation . . . [I]t is difficult to find the language that can touch that experience . . . Their behavior expressed compassion, grief, comfort and a willingness to afford assistance. I can only describe the actions of the cattle as reverential.*

Such glimpses into the unseen, unrecorded culture of the cattle that has formed up on Swona in our absence afford us insight into the true nature of an animal too often dismissed as a dim-witted, cud-chewing automaton. They give us insight into the weight afforded to death among a species we farm and slaughter on an industrial scale. If we do not see the remnants of this behavior among those more carefully tended, it is because we do not give them the chance: they have not the freedom to demonstrate it; they do not typically see out their lifespans to their natural conclusions.

And, unlike on a farm, the bodies of the dead remain where they fall. While I am on the island, I come across two carcasses at advanced stages of decomposition, each announcing themselves loudly on the air, a stench that sears my nostrils and coats the back of my throat. In crime novels, the "scent of death" is usually described as sweet; on Swona I discover it instead to be thick, miasmic, unmistakably of flesh. It is grossly intimate, utterly unlike anything else of the island—the mildewed must of the wet plasterwork, the vegetative stink of manure or rotting seaweed, the reek of the guano that rises in ragged ridgelines below favored perches.

The first time I catch the scent, I can't place it, but my body reacts immediately and instinctively with horror. It is with an extreme dread that I round the corner of the barn to find a bovine cadaver stretched out across a stone slab: flesh melting away to a rank gray liquid, the belly and pelvic region spongey and fibrous. The bones of the chest and spine are revealed afresh, blond and sculptural, primitive-looking; the front legs broken off above the hooves lending a dinosauric appearance; the jaws long and narrow, almost beak-like where it comes to a narrow point at the muzzle, and hanging open.

For many months after a death, the cattle will visit and revisit the bodies of their fallen—the way elephants are said to do in the African savannah. They sniff them, touch them. As months pass and the flesh slips away and the skeletons are laid bare, they will unintentionally step upon them and break them apart. In this way, over time, the bones will be ground down and returned to the earth. An ancient ritual of the cattle that otherwise we might never see.

To me, the question the cattle of Swona prompt is this: can a domesticated animal ever become wild once again? To answer it, first it requires us to know what it means to be *domesticated*.

Domestication is a relationship between humans and animals that grows up over generations. This goes beyond mere *taming*—a beautiful, complex, fascinating process in itself, at the level of the individual (the tempting, the wooing, the working at a creature's resistance like a knot; the breaking down of fear, of reluctance, of defiance). Domestication is how tameness, or something like it, becomes ingrained in the species' very soul, through the selective breeding of an animal that impacts its future forms.

Selective breeding often centers upon a physical trait—a *meatiness*, for example, among beef cattle or chickens; speed among racehorses; pelt thickness or richness of color among mink; the sweetness or size of fruit among plants. That fast horses bred with fast horses might produce yet faster horses is simple enough a concept to grasp; more interesting is what happens when one selects by temperament. By selecting for docility and friendliness, a certain predisposition toward tameness might be made an inherited trait. In return, humans bond with and protect their familiars: we feed them and encourage them to breed. In this way, friendliness brings selective advantage—and becomes ever more concentrated.

With this inherited amiability often come other traits too, which may seem on the surface to be entirely unrelated, or only intuitively so. Darwin himself suspected this: "Not a single domestic animal can be named," he noted in his *Origin of Species*, "which has not in some country drooping ears." His hazarded explanation—that these floppy ears were "due to the disuse of the muscles of the ear, from the animals being not much alarmed by danger"—now seems fatuous, but in his initial observation he was onto something.

In the late 1950s, Soviet scientist Dmitri K. Belyaev began an experiment: he took two groups of silver foxes and bred them selectively. One group he selected for tameness—considered to comprise curiosity toward humans and the disinclination to bite—and the other for the opposite: fearful behavior and aggression. Only the foxes in the top 20 percent in each group were then allowed to breed; the process was then repeated with their offspring.

As the generations mounted up, other changes came over the group selected for friendliness. Not only were they more interested in humans, but they began to display behavior more commonly associated with pet dogs: babyish or obsequious behaviors such as licking the hands of their keepers and wagging their tails, and—more strangely—the appearance of visual markers. By the fortieth generation, many of the fox pups were keeping their floppy ears longer, had curly tails, short legs, white patches (piebaldism), and unusually pale pelts (dilution). They also showed a change in breeding behavior—coming to sexual maturity earlier and breeding year-round, a quality that fur farms had previously attempted to select for and failed. Together, this ragbag assortment of qualities appear in many domesticated strains and are slung together under the heading of "domestication syndrome." An analogous suite of domestication-related traits is found among food crops also: larger (but fewer) grains, the loss of natural seed-dispersal methods, synchronized flowering, and the decrease of bitterness in taste.

The depth of the silver fox's domestication should not be overstated; seventy years of selective breeding falls a long way behind the dog's fifteen thousand. The foxes remain unpredictable, pissing profusely and profanely indoors and out, and being altogether somewhat unmanageable and unsuitable as house pets. But the extent of their transformation shows the impact human selection can make upon the nature of our fellow creatures. From wild-eyed snapper and screamer in the night to wet-nosed sniffer at trouser legs—all this in the space of a few dozen generations.

This is what it means to be domesticated.

What then, does it mean to be *wild*?

This, perhaps, is a more mutable quality—one whose definition depends upon who is naming it. If by *wild* we mean to survive without human input—truly apart, not scavenging of the scraps of human society—then the cattle of Swona are wild already. If by *wild* we mean wild in behavior, antipathetic toward humans, unpredictable, uncontrolled, and uncontrollable, then they are already wild. But if by *wild* we mean untouched by man—never domesticated, never tamed, never "tarnished" in that way—then perhaps they can never be wild again.

Certainly the cattle of Swona are *feral*, that is: creatures that were once domesticated reverted to a "wild state." There are feral animals of all kinds around the world. Herds of feral horses (mustangs, brumbies, criollos, and so on) move freely across the rangelands of Australia and the Americas. Feral pigeons hobble knock-kneed in city streets. Feral swine cause chaos in the rural American South. Feral dogs snap at heels and scavenge for scraps in urban badlands. This word *feral* has complicated connotations—that of viciousness, unkemptness. A certain *impurity* of nature.

But is there a point—now, or at some time in the future—at which the balance will tip? When the demands of their abandoned state has unpicked the thread of domestication so completely that a once-domesticated species might be considered truly wild once more?

In the case of the cow, this thread must be teased out from a long way back. All cattle, all 1.4 billion of them worldwide, are thought to descend from a herd of maybe eighty animals domesticated from the aurochs, or wild ox, over ten thousand years ago. The aurochs was a gargantuan creature in the mold of ancient megafauna, growing to six feet at the shoulders, and crowned with a set of great arching horns of almost comical proportions. In a commentary on his Gallic wars, Julius Caesar described them as "little inferior to elephants in size ... Great is their strength and great their swiftness; nor do they spare man or beast when they have caught sight of them." They "cannot be habituated to man, or made tractable," he continued, "not even when young." Wild, then, in its very purest sense.

Yet already by Caesar's lifetime the aurochs were in freefall. Though they once ranged across much of Eurasia and north Africa, by the Middle Ages their habitat had shrunk to eastern Europe—where their hunting was limited first to noblemen and then to royalty in a bid to halt the decline—until finally it shrank to only the Jaktorów forest in Poland. In 1627, that was where the last known aurochs, a female, died: one of the world's first recorded extinctions.

That was it for the wild ox. Or perhaps not quite. The resurrection of the aurochs has become a scientific obsession over subsequent generations, a toy that cannot be put down. The thinking behind it is this: though the last aurochs is long passed on, does enough of its genetic material survive in the domestic cow to allow it to be pieced back together? Can, through some process of biological transubstantiation, the aurochs be reborn from the body and the blood of the humble cow?

Efforts in the early twentieth century were spearheaded by two German brothers, Heinz and Lutz Heck, whose parallel attempts to re-create the aurochs by a process of "back breeding"—that is, selectively breeding existing domestic cattle with aurochs-like characteristics (such as horn type, temperament, or color)—are somewhat notorious due to their connections to the Nazi party. Each project harked back to some "pure" historic archetype—the Hecks and the aurochs, or ur-cow, the Nazis and their Aryan *volk*—and shared a vision of a primordial Germanic landscape. The younger brother, Heinz, was a somewhat ambivalent participant in the Nazi movement, but Lutz explicitly and enthusiastically linked his back-breeding project with Nazi plans to expand into ("reclaim") eastern Europe, reframing the Lebensraum project as one of ecological restoration, and giving scientific legitimization to broader Nazi theories of eugenics. In this, Lutz gained the personal backing of, among others, Hermann Göring.

Lutz concentrated upon temperament, using the ferocious Spanish and French fighting cattle as a starting point. Heinz took a looser approach, based more on appearance, taking an assortment of the steppe cattle of eastern Europe, Highland cattle from Scotland, and assorted others, throwing them all "into the one pot." Either way, both soon claimed success. As Lutz declared in 1939: "The extinct aurochs has arisen again as a wild German species in the Third Reich."

Lutz's monsters did not live long; they perished in the bombing of the Berlin zoo. (His back-bred "wild horses" and imported wisents, European bison, let loose in the forests of Poland, were killed too during the German retreat, on Göring's command.) But Heinz's persist today as the "Heck" breed of cattle, which have found favor with some proponents of rewilding—who believe that by reintroducing ancient species—or their closest substitutes—we might return our land to a more ancient form.

Heck cattle were chosen for the Netherlands' Oostvaardersplassen, for example: a controversial nineteen-square-mile reserve hailed as a "new wilderness" by supporters, but has, in the past, been decried as "an Auschwitz for animals" by critics after several instances when hundreds of animals starved behind its fences during harsh winters. (Animals considered unlikely to survive the season are now preemptively shot.) Other ecologists remain dissatisfied by the diminutive Heck, and are making fresh attempts to revive the aurochs. One project, financed by Rewilding Europe, has herds of an aurochs-like type of cattle called the "Tauros." The Tauros, claim the researchers, is "born to be wild."

Is it? Perhaps the infamy of the Heck cattle breeding project has prejudiced me against this way of thinking, but I do worry that, in pining for the landscapes and the wildlife of the distant past— ones that may never have existed in quite the way we envisage— we risk creating false idols. Carefully constructed selective breeding projects to produce "wild" animals seem, on the face of it, to be self-defeating. What could be more artificial than selectively breeding animals based on superficial traits—horn shape, height at the shoulders, shade of brown—attributed to an animal as much myth as reality? Even those bred for aggression are carefully controlled: any progeny showing *too* much a tendency for bloodshed will be shot—lest Frankenstein's monsters run amok.

Others argue that Heck cattle and those like them—those bred for toughness and independence—can serve an important ecological role in a landscape being actively managed for biodiversity: their browsing and grazing habits inhibit the succession of closed-canopy forestry, allowing for a more complex ecosystem in a smaller area. Cattle *have* been used to good effect for this purpose at sites like Knepp, in England, and elsewhere.

Still, it's complicated, and the role of semidomesticated animals as forces of rewilding is somewhat ethically unresolved. Think of Robert Elliot's classic essay, "Faking Nature," a treatise on the ethics of ecological "restoration" written in 1997. In it, he argues that there is some intrinsic value in untouched nature that he compares to a work of art. The pristine landscape is like a masterpiece, he says; a "restored" one like a forgery. Provenance is all—the "special kind of continuity with the past"—and therefore no matter how skilled the forgery, the value is lost.

Perhaps there is some commonality here with the project to forge an aurochs-copy so convincing one might never guess the truth. But is there not something more authentic, more honest, in the humble Swona cattle—visually indistinguishable from any normal herd, but whose domestic past has long passed from living memory among its members?

For something profound is at work. It is obvious here on Swona, where the stink of the dead rises from the ruins, that it is not only the behavior of the animals that is in flux. Ten or more generations have now passed in this new kingdom of the cattle. There have been many deaths and many births. The process we call natural selection is coming back into play—perhaps, for this population, for the first time in ten thousand years.

They depend now upon their own collective wisdom in breeding, and face many dangers of little concern to their husbanded cousins. Very quickly, certain traits can become selected for. The ability to thrive on little food, for one. Ease of calving. (Normally, with pregnant cows, around half of all first-time mothers would require assistance in birth—which on Swona, a world away from the calving sheds where exhausted farmers patrol with calving straps or chains, spells painful death.) Among males, dominance and aggression. There is a term for this: "reverse evolution." That is, the reversion to an ancestral form after a return to ancestral living conditions.

There are two methods by which reverse evolution might take place—the first depends on something you might call genetic memory: fragments of the species' evolutionary history survive in a scattered form in the DNA of its present-day members. Obsolete genes hang around long after they have fallen out of use; if past conditions reoccur, they may again offer an advantage and become reestablished. In this way, the ghosts of past breeding decisions can be seen moving through the herd by their coloring; the black cows carry more Aberdeen Angus heritage, the browns carry more of the Shorthorn. White belly markings suggest the past mating of a Shorthorn bull to an Angus cow. These are genetic memories of domestication. But there are older memories in there too. The Swona cattle have reverted to calving only in spring, a trait common in wild animals.

However, such memory is partial. As time goes on, more and more obsolete genes are completely "forgotten." Experiments on fruit flies suggest that genetic memories might continue for anywhere between two hundred generations and a thousand—after which any lost trait must be evolved again from scratch.

And, as the animal scientist Stephen Hall is at pains to remind me, evolution moves slowly—and a population as tiny as that of Swona's cattle can expect to be tossed high on the waves of chance: one bad spring might wipe out all calves that year, except all those born unusually late—thus inadvertently (and wrongly) selecting for late birth, for example. Inbreeding too can allow strange and damaging genetic characteristics to become widespread—although in certain circumstances, when combined with natural selection, it can also have the effect of purging harmful recessive genes from the genetic pool.

So, regressive, creative, and random forces come into play; together they work to readapt a domestic species to wildness, a process that has been dubbed "de-domestication." Genetic drift can happen quickly. In 1999, less than thirty years after their abandonment and isolation, the Swona cattle were granted a new entry in the *World Dictionary of Livestock Breeds*, the first such addition in more than a century. (The rabbits on Swona, descended from pets set loose in the 1920s, are already well on their way to regression to their ancestral type, being less thoroughly domesticated. Though initially black and white, rabbits on the island now appear brown, like their wild forebears.)

It is extremely unlikely, even given ten thousand more years, that these de-domesticating cattle will revert to their ancestral aurochs form. After perhaps 2,000 or 2,500 generations in domestication, a huge amount of aurochs genes have been lost to history. We will not return in two centuries to find Swona teaming with wild-eyed, thick-hided beasts standing six feet to the shoulder. (Indeed, most island mammal populations tend to shrink. Dwarf elephants roamed Malta and Cyprus in the long-ago past. A herd of two thousand feral cattle on Amsterdam Island, a tiny volcanic outcrop in the Indian Ocean, lost a quarter of their body mass in the time between the introduction of the original five individuals in 1871, and 2010, when they were all shot dead by conservationists.)

But at some point, any question over the "authenticity" of the Swona cattle—the independence of their existence—will fall away. After a certain amount of time, feral animals become wild beasts, no matter their domesticated past. By then they will be evolutionary works of art all their own.

Later I head out onto the open expanse of the island's north headland. The ground beyond the dry stone wall is deeply pocked by hooves; where the grass has worn thin it lies cratered and dusty as the moon. The salt breeze has dried and hardened the earth, making it hard to walk. I stagger across it, lifted by the force of the buffeting wind that blows full in my face. I raise my feet high and place them carefully down again, a spacewalker moving into unknown territory.

I have no particular goal in mind, but find myself drawn toward a dry stone cairn that stares over the precipice at the northernmost point, roughly the same height and shape of a human figure. Fulmars and kittiwakes fly up from their nests in the cliffs as I pass, swirling and churning in the air, the oystercatchers still sounding their piercing alarms. The ground is grassed over here and glinting with golden wildflowers—bird's-foot trefoil, buttercup, silverweed, coltsfoot—while along the cliff's edge the sea thrift is dying back: once cotton-candy pink, its outer stems are now fading and gray. Here and there neat stacks of flagstones, quarried long ago and now left to the elements, are splashed in clashing lichens—cream and mint and that radiant, almost fluorescent, marigold.

The whole island feels to be growing incensed by my presence. A new and startling sound starts up: a juddering moan, rising in pitch and frequency, then dying away again. There's something glitchy about it, like radio interference. It takes several passes before I even understand that the sound must be emanating from a bird. Finally, after twenty more, I put a name to it: a snipe and what they call "drumming"—a loud, vibrato buzz produced through its tail feathers whose timbre shifts in a dopplered curve as the bird passes overhead. It's getting to me. I shake my head of the noise, push on, feeling harried and strangely panicky.

The air around and above my head is alive and threatening, as seafowl dive at my head as close as they dare before whirling away again. The ferocity of the birds is increasing in degrees with every step. Then, suddenly, the sky is filled with a new kind of bird. Tiny, pale-chested, with blood-red beaks and black masks pulled over their eyes. Arctic terns. They materialize over my shoulders, beating their wings at me, their forked tails honed to fine points, clicking loud as gunfire, punctuating their bombardment with piercing screams.

One darts at my face and I flinch away, and feel it flip my hair as it brushes past. I raise an arm protectively and feel wing beats, the scratch of small, sharp claws. I catch a glimpse: scores of these elegant, razor-edged creatures hover malevolently in the corners of my vision. At some point, I cannot advance any farther. I have met total resistance.

I beat a hasty retreat. Their viciousness has caught me unawares. Images from nature documentaries come to mind: eagles being harassed by crows midair; lions repelled by wildebeest. Apex predators brought low by sheer force of numbers, by collective outrage. Still under siege, I back from the terns, moving as fast as I can without tripping on the pocked ground. The wall of aggression retreats, although the oystercatchers are close at my heels, piping their warnings for all to hear.

Thoroughly shaken, I keep on the move, heading south and into rolling landscape in the island's interior. But every time I flee one creature's territory, I move unwittingly into another's. Past Keefra Hill, the cattle are standing in watch, alert and suspicious. Soon I pass into the land of the great skuas. They are stocky and aggressive, known to physically attack, vomiting or spitting at interlopers. Here, each heavyset bird appears to occupy a small circle of ground, rearing threateningly toward me as I enter its arena, then falling back as if chained as I pass on into the next.

Finally I cede any claim to solid ground and climb out along the rocky ledges of the shoreline. Not far away, a dozen fat seals bask in the watery sunlight at the water's edge. They groan great guttural grunts of complaint upon my arrival, but otherwise ignore me. Fulmars patrol overhead, elegant birds outfitted in a dozen shades of gray, swooping by in long curving flights, but don't dare enter my personal space. Farther out, I see razorbills, guillemots, tiny puffins bobbing on the waves. Eider ducks paddle with their ducklings in the shallows. For the first time in hours, I relax. I try to regain my composure. I remind myself: if one goes in search of nature in its wildest forms, you shouldn't expect it will be pleased to see you.

On my way back I pass the ruined barn, where the old cow is laid out across the flags as if for a post-mortem. Then Rose Cottage. I avert my eyes. In British hospitals, that phrase is a euphemism: nurses ask porters to transport the recently deceased to "Rose Cottage" instead of to the mortuary, to avoid distressing other patients. I try not to think of that body, prone on the parlor floor and sealed up tight.

———

When I take to my bed the sky outside is still and pale and primrose. It's nearly solstice.

In an upstairs bedroom I have found an old camp bed—or the metal bones of one, rusted in place and strung with springs. I place my camp mat on it, brush off the mice droppings, and slide under my blanket, with my jacket as pillow.

Just as I get settled, I hear a movement—very real, and very definitely inside the house. It sounds like a man running fast across a wooden floor. It's not clear where it's coming from.

I slide from my bed and stand stock-still in the doorway, listening. Nothing. I creep along the landing and take the stairs back down, stepping as softly as I can on each tread, and force myself to peer around the door into the kitchen. Nothing.

Then the sound again: running. It's definitely upstairs. I think immediately of the disappearing tent, the footprints, their source. It occurs to me that someone might have been on this island the whole time. I feel a wave of nausea. The prospect of being alone doesn't bother me. It's the thought of being *alone with*.

I return upstairs, knowing I must face it rather than live with the horror of uncertainty. I peer into each bedroom in turn: nothing. There seems no answer. Still, I relax. I am afraid of many, many things, but at this moment I realize I am not afraid of ghosts.

I stand there for ten long minutes, breathing shallowly through my nose. Finally, the sound comes again, louder this time. I am close. I see nothing, but I approach the noise until I'm sure: it's in the ceiling, behind the wooden paneling. I catch my breath, tighten my throat, and knock hard on the wood, near the source. It ignores me.

There are no rats on this island, this much I know. But I picture the roof: those slipped slates. A bird, I think, between roof and gable ceiling. I hear it again, traveling from above my head through to the bedroom where I set up camp, moving fast with a heavy, rapid footfall. I follow it, and climb back into bed as it fusses and thumps.

All night the unseen creature moves in the space above my head. It runs first one way, and then the next, centimeters away. Never before have I been so utterly aware of how the wild exists just below the surface. Even in domestic animals. Even in houses. I drift in and out of sleep, as if resting only half my brain at once, judging the passage of time by the shifting tones of the sky, the constellations in their unhurried revolution.

I come to around dawn, when I take a flask of tea outside to drink on the front step. It's blustery but clear, the whole land lit with a flat yellow light, the grass inlaid with flecks of white and the palest of pinks: daisies and ragged robin and eyebright. Farther out, where the hill rises behind the last of the old houses, the broidered sward is rubbed through like an old toy loved bare, its stuffing come through as the downy fuzz of bog cotton. Dockens grow in tangles at the feet of old walls.

While I wait, the cattle appear over the brow of the hill, coming down to graze in a grassy hollow by the rocky strand. Their winter coats are on their way out. Some are shedding and bedraggled still, their hair standing in riffles and cowlicks, but others are already as glossy and as sleek as chestnuts.

They move as one animal: comfortable in one another's company, taking long, relaxed strides. They don't see me. It doesn't occur to them to look.

PART FOUR

ENDGAME

11

REVELATION

Plymouth, Montserrat

I t was a day like any other day: July 18, 1995. The "emerald isle" of Montserrat, a British territory in the Caribbean, awoke to clear skies, another steamy summer's morning. But then: a strange thing. In the island's capital, Plymouth, white powder began to sift silently down upon the streets. It shimmered down over everything, like pollen.

Castle Peak, a weathered nearby mountain of igneous rock shrouded in dense forest, appeared to have sprung a leak. A ragged plume of steam issued from a crack in its base, and rose high into the air before dispersing over the town as ash. Curious townspeople hiked the mountain paths to see it, pushing through ferns and lush rainforest. It roared, they said, like a jet engine.

It was, of course, alarming. But the vent wasn't violent enough to be truly frightening—it was more an object of curiosity. They were used to the hot springs that had bubbled out from underneath the peaks for as long as anyone could remember. But as the days and weeks went by, activity at the vent increased. Sulfurous fumes blew down the mountain in a mist, obscuring the view of the peaks from the town. The fumes had a strong smell, not exactly unpleasant—like that of a just-struck match—but sometimes grew powerful enough to burn the nostrils.

And the ground beneath their feet—it began to quiver. Not to *quake*, exactly, but a slight tremor; just enough to make one doubt one's senses, to set one's nerves on edge. At night, the residents of the slopes of the Soufrière Hills lay in bed listening to a faint grating sound emanating from deep below, like heavy machinery grinding into action.

Several more vents cracked open, and from them spewed rivers of a thick gray mud, which flowed down through the forest and out to sea. The ash grew in quantity, becoming thick like smoke, sometimes rising twenty thousand feet before falling back over the island as black snow. Unease began to grow.

If this was an emergency, it was a slow-moving one. Weeks turned into months in a twilit state, the ragged peaks of the Soufrière Hills cloaked in thick black clouds that sometimes rolled down the mountainsides and into the town, blocking out the sun and turning day into night for minutes at a time. Then it would clear, and the townspeople would sweep up the precipitate, and wait. Like a pan coming to a boil, heating slowly but inexorably, it wasn't clear when they should call it, and get out.

The experts weren't sure either, although they tended toward caution. By the following summer, Plymouth had been officially evacuated for the second time. But the residents living in squalor in church halls and gymnasiums were growing restless. Many left, for neighboring islands or taking the paltry sum on offer to resettle in the UK. Others still—especially farmers, fretting over their animals—opted to return to their homes, against official advice.

One hot August day, just after noon, pastor and youth worker David Lea was in the empty town inspecting it for ash damage. His beeper went off. It was his wife, Clover. When he called her back, she sounded flustered. It's coming, she said. Get out of there.

David jumped in his car, but as he roared up the road out of town he saw a huge black mass moving fast out of the corner of his eye. It moved, he thought, like a dragon might. And as it closed in on him, he came into what sounded like a hailstorm—hard pebbles began to tap and then to dash themselves against his windshield—tiny rocks of what he soon realized was ash. Then came the larger stones. He accelerated, speeding toward safety. Less than a minute later, the car was overcome by darkness.

Blackout. As the scriptures say, of the ninth plague upon the Egyptians, a darkness that could be felt. David hit the brakes and slowed to a crawl. The temperature inside the car began to rise.

Then came the lightning. It flashed in all directions, from inside the cloud, seemingly emanating from the darkness itself. And the thunder! Constant thunder. David's clothes were soaked with sweat. He hunched over the wheel, driving at two or three miles per hour, wipers flapping uselessly against the bombardment.

Then: there—up ahead, he saw the blinking taillights of another car. He cried his thanks aloud. For twelve long minutes they drove bumper to bumper up the track and out of town.

For a time, he said, it seemed like it might never end. But as they reached the bridge across the River Belham, finally he saw a glimmer of daylight.

When he emerged from the cloud, he found the landscape changed beyond comprehension. It was ill-formed and dull, as if someone had dialed down the color saturation. A rain of mud had cascaded from the sky and lay two inches thick over everything—rendering cars, houses, trees, roads, animals, and people unrecognizable. The ground was slick with it, his car sliding backward through the muck.

But he was safe, he was safe, he was safe. Behind him, the volcano roared. It was becoming clear that there would be no going back to the way things were before.

Another year later—a year of rumbling and blackouts and mounting fear—the volcano finally wreaked the destruction it had promised. On June 25, 1997, a massive eruption saw 150 million cubic feet of magma ejected from the Soufrière Hills volcano in the form of pyroclastic flows: roiling avalanches of superheated rock, steam, ash, and fumes that coursed down the mountainside in a state somewhere between a liquid and a gas.

If lava is deadly in its liquid incandescence, pyroclastic flows are something else. They rush the land as a devastating force, moving faster than a speeding car—downward, laterally, over water, even sometimes uphill. It is impossible to outrun them. Should they claim you, they will boil the blood in your veins, crack open your skull, vitrify your brain, vaporize the flesh from your bones.

In Montserrat, the flows rapidly subsumed one and a half square miles of land, and killed nineteen residents who had returned to the area against official advice. Those who died, died instantly, burned by material of up to 750°F, or suffocated by the choking ash.

From outside, the flows had been almost silent—spilling down the slopes and consuming its unsuspecting victims without warning. The huge black clouds were all-engulfing and inside was noise and chaos—winds rushing as in a hurricane, the black flow pitching like a stormy sea, fires whirling, gyratory, explosions booming somewhere out of sight.

Search-and-rescue parties helicoptering in afterward found a scene of utter devastation: ash pulled taut over everything; vegetation scorched from the earth, plow furrows baked hard as clay; utility poles awry and hung with bare copper wires where their insulation had melted off. Glass panes shattered or distorted like plastic. China ornaments melted on their sills. "It was like the whole place had been bombed," one resident recalled later. There was an acrid smell of burning tires, and fine white ash puddled and splashed like water under their feet.

In narrow clefts, pumice built up in huge, foaming mounds like bubbles in a bath, while boulders the size of cars, aglow with heat, rolled like marbles, sending up fireworks when they collided. Buildings that escaped the flows were baked where they stood: aluminum shutters warped and twisted, wooden doors and frames burst spontaneously into flames, glassware in the cupboards softened and reset, bowed into strange new shapes.

One family was saved by the Methodist church behind their house, which shielded them from the brunt of the flows. The parents ran into an interior room, pulling their four-year-old daughter behind them; all but her trailing arm had made it safely inside when the first surge hit the house. She was burned from shoulder to elbow, but otherwise escaped unharmed. Another survivor, who lost his wife to the blast, only escaped by running over red-hot ashes in his bare feet; his toes were later amputated. Another escaped the pyroclastic onslaught in his car, the treads of his tires on fire. Panicking inside the vehicle, in the blasting heat of an oven, he turned on the fan only to receive a faceful of scorching ash. Nine bodies were never recovered and may never be—they lie buried under deposits several yards thick.

A few weeks later, as the islanders were still grieving the losses, the volcano roared into action yet again. This time, Plymouth lay directly in the path of the flows. Early on August 4th, they saw over the ridgeline the awful glow of their homes burning, the licking of the highest flames rising hundreds of feet into the air. By the morning of the 8th, the capital lay under five feet of ash.

It was a waking nightmare, all of it. Having never known they lived at the foot of a volcano, they had seen in the space of a few months almost everything they held dear smothered by ash, or set alight, or locked away in an exclusion zone that now was thrown up around the southern two-thirds of the island.

Yet, through the grief and the terror, the whole experience had been laced with something like astonishment at the terrible majesty of the thing—the awesome power of the forces bursting from the Earth beneath their feet. "It was something, man," said one survivor, in a tone approaching wonder. "It was really something to see . . . the most beautiful thing." It was "like burning hell down there," said another. "If anybody had told me this would happen to me, I would never believe it."

They had seen Armageddon and lived; experienced the apocalypse in surround sound, felt it singe their skin. Impossible, perhaps, when faced with such horror not to find oneself astounded by the forces at work: the sheer wanton destruction that the Earth was capable of. (One can compare, perhaps, that transcendental wash of fear and reverence that must have seized David Johnston, the young volcanologist on duty at an observatory overlooking Oregon's Mount St. Helens on the morning of May 18, 1980, when he became the first witness to the dome's collapse—and, grabbing his radio, voice cracking, shouted, "Vancouver! Vancouver! This is it!" only seconds before he was overcome by a wall of fire. In Hawaii, traditional beliefs hold that lava is the physical embodiment of Pele, the destructive, ravenous goddess of fire. Any attempts to divert the flow—even to protect against death—are therefore blasphemous.)

After his experience in the ash cloud, David Lea had found himself drawn to spend more and more time courting the volcano.

He filmed its activities, initially on a home video camera, but then on increasingly expensive, professional-standard equipment. He took more and more risks. *Calculated risks*, he felt, but risks all the same. He rented a house high on a ridge overlooking the Soufrière Hills, leaving Clover with the children in their safe clifftop home on the coast. He bought a fire-resistant suit, then breathing apparatus used by scuba divers, then a "bomb-shelter" made out of an old concrete cistern, with a hinged steel door that might be swung shut behind him in an emergency.

It was madness, he knew. But he couldn't resist. A religious man, he felt the presence of God in the natural wonders unfolding in front of him, in the "divine appointments" that led him to witness them. Shrouded and gray during daylight hours, the volcano was seen to its best advantage at night when it was lit by an infernal glow, incandescent rocks spilling from its mouth. Once, hiking a dirt road into the south of the island—rendered impassable—an incredible roar rent the air; pressing himself against the rock face, he waited for impact. None came. After a few minutes, he staggered forward to the overlook, from where he stared out across a river of fire flowing from the dome through the valley and into the sea. It was intoxicating, the raw thrill of it, the terrible beauty.

When the volcano lapsed into silence—repose, as volcanologists call it, as if in gentle slumber—David and his son Sunny would venture into the ghost towns and villages left empty in the wake of the eruptions, and find within a new category of spectacle. Extreme temperatures have strange, alchemical effects. David and

Sunny found trees and bushes subjected to such intense heat they had melted into tar. They found houses blasted so cleanly from their foundations that only their tiled floors remained. They crept inside buildings that had escaped devastation and found dining tables set carefully for dinner, the whole scene shrouded in a perfect pale dust sheet made of ash, the windows melted, dripping from their frames.

And there were other times too, unexpected pleasures of a purer, less macabre, kind. Once, having picked up warning of an imminent blast on his scanner radio, he stopped at the side of the road and turned to face the volcano. It was raining. Visibility was poor. A rainbow arced cleanly from the clouds. Then came the explosion: so powerful that a plume of ash and steam shot upward in a column twenty-five thousand feet high, sucking with it both the rain clouds and the rainbow and leaving him standing, suddenly, in a bright, clear summer's day. *Mind blowing*, he said later. After eruptions like that, ash would snow down again, tons of it, six inches at a time: a soft, dry snow that wouldn't melt. Once, it blew so high into the atmosphere and the ash was carried so far out to sea, that they were spared even a mote of dust, as if the explosion had never happened.

But if the volcano of Montserrat was impressive, larger eruptions have the capacity to mesmerize the whole world. In 1888, the Royal Society published *The Eruption of Krakatoa, and Subsequent Phenomena*, a collection of fantastical reports following an 1883 eruption so massive it had been heard 3,300 miles away on Mauritius. In the weeks that followed, the Royal Society received "voluminous correspondence" from around the globe describing glowing skies and cosmic happenings they could not help but associate with the blast that had claimed thirty-six thousand lives.

In Honolulu, a thin film was seen high in the atmosphere, "perfectly transparent" but visibly rippling, causing a corona of "faint crimson hue" to appear around the sun. In England were seen opalescent skies, a moiré of rainbow colors; as in Norway, where they are thought to have inspired the lurid backdrop to Edvard Munch's *The Scream*. In India, the sky grew green, and took on a mottled, smoky aspect. Sometimes the sun itself was green; in San Salvador, the moon appeared as a "crescent horn deep green in the midst of an immense crimson curtain." In Pennsylvania, one newspaper went as far as to claim: "an immense American flag, composed of the national colors, stood out in bold relief high in the heavens." Patriotic fervor apart, all reports were of one voice: the skies were extraordinary. Krakatau,* it would seem, was the cause.

We now know why: sulfate aerosols—tiny droplets of sulfuric acid, were lofted so high by explosive eruptions that they became trapped in the upper layers of the atmosphere. Similar celestial phenomena arose in the aftermath of the eruption of Tambora in Indonesia in 1815, the most powerful volcanic blast in recorded history—ten times bigger even than Krakatau. Of an estimated twelve thousand people then living in the district surrounding Tambora, only twenty-six survived.

---

*Though originally rendered in English as "Krakatoa," Krakatau is now considered the correct spelling.

Across the world, the sky bore witness—reflected in the luminescent skies of J. M. W. Turner's paintings of that era. The sulfate particles swirling through the upper atmosphere also had profound impact on global weather systems for three or four years afterward. In 1816, freakishly low temperatures and high rainfall resulted in catastrophic crop failures across the entire Northern Hemisphere. In Yunnan, China, crop after crop of rice failed and many resorted to eating white clay; in Ireland, an estimated one hundred thousand died after potato crops failed. Food riots broke out across central Europe; false prophets caused panic with forecasts of mass destruction. In the United States, it snowed on Independence Day celebrations as far south as Virginia.

Similar crises of famine and civil unrest are thought to have broken out in the years following the eruption of yet another Indonesian volcano, Samalas, in 1257, which released around a hundred million tons of sulfur dioxide. Reconstructions indicate this could have induced global cooling by up to 10.4°F for a period of four or five years (though the true extent is a point of rigorous debate). Certainly, the impact of the eruption reverberated around the globe: the following January, English monk Matthew Paris recorded "such unendurable cold, that it bound up the face of the earth . . . suspended all cultivation, and killed the young of the cattle," and, by summer, starvation: "dead bodies . . . swollen and livid, lying by fives and sixe's in pigsties, on dunghills, and in the muddy street." (A mass grave holding in excess of ten thousand corpses dating from this period was unearthed in Spitalfields, London, in the 1990s.) Glaciers and ice caps expanded.

What these massive eruptions show us is the dynamism of the Earth's climate—and the relative ease by which normal functioning might be knocked off balance. In this way, a major eruption might create a disaster with two tails: the immediate crisis, and then the longer tail of climatological chaos.

Eruptions on Tambora's scale—technically classed as a 7 on the Volcanic Explosivity Index, which travels from 1 ("gentle") through 3 ("catastrophic") and 5 ("paroxysmistic") up to 8 ("mega-colossal")—are forecast to take place twice every thousand years, or thereabouts. Level 8 eruptions—"supervolcanoes," like the one under Yellowstone National Park—are thought to occur somewhere in the world every 100,000 years or so, but possibly as often as once in every 30,000. Supervolcanoes pose true existential threats; when Indonesia's Toba erupted 74,000 years ago, it is believed by some to have caused a volcanic winter of such ferocity that humans were almost entirely wiped out—the entire global population falling to between only three thousand and ten thousand individuals.

And it is a supervolcano—on the grandest possible scale—that is now believed to have triggered the single largest extinction event the world has ever seen, the Permian extinction. Around 252 million years ago, massive eruptions in what is now Siberia propelled as much as 720,000 cubic miles of ash into the air—and such enormous quantities of greenhouse gases (an estimated 1,200 billion tons of methane and 4,000 billion tons of sulfur dioxide) that the global temperature rose by around 18°F. Afterward, forests rotted where they fell, oceans stagnated, acid rain scoured the surface of the earth. During this period, nearly all life on Earth collapsed: more than 95 percent of marine species and three-quarters of land species were eradicated, including most synapsids, the otherworldly creatures that then dominated the planet—leaving a vacuum in which dinosaurs would later come to the fore.

Should a supervolcano erupt again—as Yellowstone, roughly speaking, is due to do—it would be the greatest disaster civilization has ever seen. Millions would be killed during the immediate blast. An entire continent would be blanketed in ash, turning day into night, poisoning water, and devastating global agriculture for years. Temperatures might plunge by 32°F for a decade or more.

Should there ever again be climatic upheaval even remotely approaching that of the end-Permian: lights out. It would almost certainly spell the end of the age of humans, the end of the age of mammals.

Twenty-two years after its initial entombment, I enter Plymouth—or what remains of it—in the company of Sunny Lea. A police escort is waiting for us at the checkpoint to the exclusion zone that surrounds the former capital, and hangs back while Sunny radios in to the volcano observatory two and a half miles away in a place called Hope.

"Sierra Lima requesting permission to enter Zone V with police presence, over."

"Go ahead. Obs standing by."

We roll through the gate onto a dirt track, which runs parallel to the old, ruined road now blocked off with boulders and thick vegetation. The day is clammy and humid, but a stiff breeze is blustering in off the Atlantic. The air is hazy, faintly odorous with sulfates blowing down off the mountains.

As we progress, the wraith-like forms of buildings loom up out of lush vegetation: what looks like an old warehouse, missing its roof, the rafters fallen inward; the upper floors of an office block, with floor-to-ceiling glass. The softly undulating shapes of ferns can be seen growing inside through frosted windows. Concrete walls look gouged, as if chunks have been bitten from them, baring the rusted rebar inside. Bougainvillea shivers in the breeze, its pink and papery petals draping prettily over everything like a throw.

Around a bank of what looks like a sand dune—ash, piled high during the futile recovery efforts of the past, now abandoned and colonized by grass and thorny shrubs—two huge fuel tanks rise like sea monsters through the wave. Stairs spiral around their girths, handrails fallen away. Rust streaks their blistered white paintwork. Beyond them, the ground flattens into a rubble field where a few low-rise buildings, set far apart, stand sullenly in the gravel.

This is the town center, says Sunny. We are standing forty feet above what was ground level. The isolated buildings ahead are, in fact, the top floors of what were four- and five-story buildings. Over the years, pyroclastic flows and mudslides have overcome the town in succession, enveloping the central landmarks. Here, says Sunny, drawing us to a halt and fishing in his pocket for an old postcard that shows the town clock that stands beneath our feet: a distinctive tower topped with a cupola. In the background stand elegant whitewashed buildings with red-tiled roofs and balconies. Palm trees sway in the foreground. Then: another picture, the clock half-drowned in ash, its face just peeping above the surface, as if gasping for air. Then, the whole thing went under.

The pyroclastic flows that hit Plymouth were not so different to those that swept the towns of Pompeii and Herculaneum in 79 AD, the catastrophe that claimed the lives of more than 1,500 and rendered in perpetuity the moments of their demise. Like Pompeii, most of Plymouth now lies under ash, perhaps preserving for our descendants a record of life at the turn of the twenty-first century. Only advances in seismology saved the 4,000-strong population of Plymouth from the fate of their predecessors at Pompeii, the city of the dead.

Sunny's explanation—his attempt to situate me in the city of his past—has almost the opposite effect. The descriptions of the streets below my feet are destabilizing, vertigo-inducing, as if suddenly finding myself to be standing on a clear glass floor. The disintegration of space is only emphasized by the shadowy form of the volcano, whose ragged rim rises up behind the ruined city: it is the presence that dominates all thought within the zone, and yet hides its face behind an ice-blue veil of sulfurous clouds. These make it difficult to look at: fading from view so completely that the eyes slip through it, unable to fix upon what is solid rock, or cloud, or sky.

Every few minutes, Sunny's radio buzzes with garbled crosstalk and he holds it to his ear. It's been around a decade since the last major eruption, but still we can't relax. The volcano is a known erratic, a drunken lout known to stir into destructive rages even after years of troubled sleep. Should seismic activity spike, we'll have around ninety seconds to vacate the area. We've left the car still running, pointing toward the exit.

Sunny was a child when he last walked the streets of Plymouth. What has been most strange, he says, has not been the disappearance of the familiar beneath the ashes, but the sudden appearance of all that is unfamiliar. Take the skyline. Geology usually unfolds in epochs, too slow for human perception to register—mountain ranges rising by a fraction of an inch a year, the constant, grindingly slow collision or partition of the continents—but here they witnessed an entire mountain rise up, explode, then rise again in a matter of days. Sometimes it appeared as a dome, sometimes a spire: a thin, elegant protrusion like the tip of the Matterhorn. Then, suddenly, the structure would fail, sending mudslides or superheated pyroclastic flows booming down the sides.

Now, central Plymouth appears half-dissolved, like so many sandcastles being swept away by an incoming tide. But in this case the direction of travel is out from the land, into the sea. A wealthy suburb, on higher ground to my south—visible from here but unreachable—stands as a bank of villas staring with a haunted look out to the ocean, crowded by lush green forest growing up through residential streets.

We wander inland, to the shallower edges of the flow, where

whole buildings wade from the grit, sagging with the weight of it, or keel to one side. We see rocks the size of cattle that have rolled through ground-floor rooms and rest wedged against walls, or press up against bent metal bars. A hotel has filled with a slurry of ash which has set like concrete, before its outer walls gave way to reveal this solid record of its negative space, an inadvertent reenactment of the artist Rachel Whiteread's *House*. I see the pale curves of a bowl and cistern glinting from the dirt like dinosaur bones, the toilet seat still down.

The air tastes dry and dusty. There's a sudden scrabble of movement as an iguana spots our approach and reverses speedily through a door into the nearest building—his house—his zebra-striped tail swiping an arc through the dust. At the old police station, its louvered windows have been frosted by the scouring actions of successive ash clouds. I bend to peer through the gap and find, to my surprise, the interior lush and humid, shoulder-height with ferns which crowd inside the sheltered space. A ragged curtain still hangs across an internal window.

Nearby, a supermarket has been flooded with beige ash, set smooth and powder-solid like plaster of paris to leave a dark crawl space inside. Behind it, an office has been sheltered from the deluge: shelves still stacked with fluttering paperwork, in-boxes scattered on the floor, executive chair still pulled under the desk, the whole scene dusted with icing sugar. In a cavernous storehouse beyond, rusted carts sit askew on soft piles of ash. But my eyes are drawn to a thigh-high midden of empty cans, rusted solid like sculpture. It makes me uneasy, there's something uncanny about the way it stands—a purposefulness that sets it apart from the artless chaos of elsewhere.

Yes, says Sunny. For a while, a man known locally as "Never Me" haunted the ghost town, having decided he would rather take his chances amid the ruins of the familiar than in the bedlam of the emergency shelters. Where once he had eked out a living, dollar by dollar, carrying shopping bags to cars, he returned to wander in its abandonment and, apparently, made a feast from the old stored tins. We move on—our time has lapsed, the policeman checking his watch—but this is the image that stays with me as we leave. My mind fixes on it: the forbidden thrill of it, of wandering at liberty, alone through the dead city.

Mary Shelley's *The Last Man* is an account of a twenty-first-century man called Lionel Verney, the sole survivor of a mysterious global pandemic. The progress of the disease, which spreads miasmatically through the atmosphere, has been greatly speeded by climatic upheaval. In the world of *The Last Man*, crops fail. Sea levels rise. Whole villages are carried away by floods. Doomsday cults take root among the survivors, as thousands pour north seeking sanctuary. At the end of the book we find Verney walking alone through the Appenines, sleeping in abandoned houses and derelict inns, before taking up residence in a desolate Rome, where he wanders the empty streets and considers the city's—indeed all civilization's—decline and fall.

Shelley was drawing from her experiences during that fateful "year without a summer" in 1816, when she—then Mary Godwin, age eighteen—stayed at Lake Geneva with her half-sister, her future husband Percy Bysshe Shelley, and Lord Byron, who spent their summer inside watching thunderstorms sweep the mountains while chaos reigned in the hungry villages beyond. Their own confinement was a constructive one: Byron produced the wretched apocalyptic vision "Darkness" ("The bright sun was extinguish'd . . . All earth was but one thought—and that was death"); Mary began work on her Gothic masterpiece, *Frankenstein.*

Now considered the foundational work of science fiction, *Frankenstein* serves as a parable of man's hubris and the dangers of recklessly interfering with the natural order. But both *Frankenstein* and *The Last Man*, now two centuries old, still make for unnerving reading in their prescience and their (in *Frankenstein*'s case, seemingly allegorical) mirroring of our contemporary crisis of climate, our own monstrous creation.

Seeing the global chaos—the famine, the refugee crises—that unfolded in the fallout from the eruption of Tambora, which is thought to have lowered global temperatures by an estimated 1.8°F, brings our contemporary predictions of rapid climate change into a different kind of focus: when we consider that a global warming of 2.4°F above preindustrial levels is currently considered our best-case scenario, that a rise of 3.6°F is perhaps already unavoidable, and, if little is done to curb emissions, a rise of 5.4°F by 2100 is likely.

Since Mary Shelley was a girl our planet has already warmed by nearly 2°F as a result of human activity, and in consequence we have already seen sea levels rise by eight inches, as Arctic sea ice declines and extreme weather events have grown both in frequency and in the damage they wreak.

Comparing the brief, sudden global cooling of the year without a summer with a steady global warming is not comparing like with like, but it does begin to offer us some kind of solid, real-life frame of reference. And the apocalyptic tone of the fiction and poetry it inspired does not feel so overblown when one considers that, at the time of this writing, we find ourselves in a period of acidifying oceans, erratic rainfalls, floods, droughts, wildfires, and cyclones.

Man-made climate change has sharply accelerated in recent decades; a recent study by the Stockholm Resilience Center, an independent research institute, calculated that the current rate of warming is at around 35°F per century—or 170 times that caused by natural processes. This means, the researchers wrote, that humans have now replaced astronomical and geological forces as the dominant force of change. Earth scientist Professor Will Steffen, one of the authors of the paper, commented that "while other forces operate over millions of years, we as humans are having an impact at the same strength . . . but in the timeframe of just a couple of centuries." The magnitude of man-made climate change, therefore, "looks more like a meteorite strike than a gradual change."

This book has, by and large, focused in on the silver linings: on the grass that grows in the cracks in the pavement. But it would be remiss of me not to face, head-on, the elephant: that of irreversible, catastrophic global change as a result of human actions. As Victor Frankenstein worried, of his creation: "Had I right, for my own benefit, to inflict this curse upon everlasting generations?"

The natural world is already, necessarily, adapting as well as it might to climate change. As the ecologist Chris D. Thomas has noted, a march of the world's wildlife is underway. Two-thirds of species are extending their ranges north, or onto higher ground, as local climates shift. Animals are moving toward the poles at a rate of over ten miles a decade, he observes, or the equivalent of around fifteen feet every day. "Keep this going for a few centuries and we have a new biological world order."

But there will, inevitably, be heavy losses. Not all species can move. Some may live confined to islands, or to mountain ranges. There may be no farther north to travel, or no higher peaks to climb. Ocean-dwelling species are already proving most vulnerable, being eliminated from their habitats at twice the rate of those on land. Rising sea temperatures have already seen mass bleaching of coral reefs, including at Rongelap Atoll, the source of Bikini Atoll's recolonizing larvae, raising the unedifying prospect that man-made climate change may yet prove more devastating to that environment in the long term than the dropping of a literal atomic bomb. And along the equator, that ever-widening belt of dry and barren desert lands, drained of biodiversity: their otherworldly inhabitants strange and beautiful and far between.

When considering climate change, I find one tends to swing between poles: from extreme, almost debilitating panic to something approaching ambivalence, perhaps denial. Like continental shifts, climate change moves too slowly for the human eye to perceive. This rate of change—a tenth of a degree here, or there—lulls us into a false sense of security. We let our attention shift, allow our eyes to glance to other, more immediate, concerns. But the change is constant, inexorable, gathering pace. Remember even the apocalyptic Permian extinction—the extinction event that dwarfs all others, when the future of life itself appeared in doubt—took place over a period of perhaps a hundred thousand years. Barely an instant in geological terms; but to a human observer in its midst, living through a tiny snapshot of time, it might appear nothing much was going wrong.

More urgent is the prospect of the Earth's climate spinning ever more rapidly out of control, as increasing temperatures create positive feedback loops that amplify the effects of climate change—melting ice caps lowering the planet's reflectivity, for example, and thus causing more heat to be absorbed—or worse, the risk of passing "tipping points" that cannot be reversed: the stuttering to a halt of ocean currents, or the massive release of methane held in permafrost or in frozen layers of the sea floor. Should such a calamity befall us, we may find ourselves watching climate change in real time, just as the Montserratians saw mountains rise and collapse, rise and collapse, before their very eyes.

During periods of rapid climate change, societies collapse. Civilizations fall. Past natural climate change events have shown us the fragility of our survival on this Earth—and by extension, the survival of all else that lives here with us. We are the meteor. We are the supervolcano. And it is becoming clear that there will be no going back to the way things were before.

We leave by the way we came in, exiting Plymouth along the dirt track through the checkpoint, waving goodbye to our police escort and checking in by radio to let the observatory know we have safely departed.

Immediately beyond the gate lies Zone C, an uninhabited buffer zone open only during daylight hours. The pyroclastic flows did not reach so far. The roads are lined with empty villas, grown through with trees. Their roofs are collapsing. Power lines sag over the roads, whose tarmac is scabbed and clotted, and almost overcome by the in-pressing of the rainforest.

Some people are still paying off mortgages on these properties, Sunny tells me. Just in case they can ever return. In a church, just off the road, I find bare pews empty and stained with guano. A dozen of what I tentatively identify as velvety house bats hang from the blades of a ceiling fan. They stir as I enter, twisting their heads toward me, some clambering over the bodies of their neighbors out of what looks like curiosity. They are cast in an amber tint, as the low sun streams through stained glass at the gable end in the shape of the cross. Elsewhere in the exclusion zone, a colony of five hundred Brazilian free-tailed bats have taken up residence in an abandoned house.

We pass the house Sunny spent his early years in—a red roof rising from a cloud of ferns—and his old school, invisible from the road, having been completely reclaimed by forest. "We used to do drills," he says, "in case we had to leave in a hurry." Then, one day, they did, and they never returned. He found the old buildings last year—bushwhacked his way in. The classrooms were full of bat droppings, but there was still chalk on the blackboards, swollen books lining the shelves, beakers laid out ready for experiments.

"My kids think this is normal," says Sunny. They think it's normal for two-thirds of the island to be off limits. To see their homeland divided into zones of varying degrees of risk. To them it seems that nothing much is wrong. But to Sunny, almost everything about his home is changed or gone. All his classmates have emigrated, to build new lives. "I am," he says, "the only one left."

We drive on. Mahogany trees stand wreathed with philodendrons. Egrets gather where they can in clearings in the forest. The curious, tailless agouti and frighteningly large iguanas dash into the shadows on our approach.

We end up at a hotel, long abandoned, which, from its perch on Richmond Hill, overlooks what is left of Plymouth. It is, I think, the most disturbing site of all: the hotel's elegant restaurant, with its floorboards collapsing under the weight of a deep-pile carpet of ash; the ground-floor bedrooms, flooded to waist height, the headboards jutting, tables overturned. A corridor leading to the rusted ice machine has grown through with serpentine tree roots thick as my forearm. A tiny office holds fluttering stacks of paperwork: a marketing plan marked 1996, the menu for a Grand Easter Buffet Lunch ("Call 491-2481 for Reservations"), booking records, files of receipts.

Out front, the swimming pool—once the island's most luxurious hangout—is filled to the brim with ash, which now serves as compost or mulch, beneath a dense thicket of grasses, reeds, and sapling trees. Colonized, I think, in exactly the same manner as the bings of West Lothian. Life rises from the ashes.

I step out onto the balcony overhanging the slope, and snap a view of the hotel. I'd seen photos taken from this very spot: the turquoise ripple of the pool, the smart red roof of the hotel behind, the dining room with its flipping fans and bar brimming with cocktails. Toward the back—where ferns now spill—a small stage for the band.

In *The Last Man*, Lionel Verney takes up his final residence in a Roman palace. As well you might, as the last survivor of all humanity. But this, I find myself thinking, is more in keeping with J. G. Ballard's *The Drowned World*, in which the hero, or anti-hero, Dr. Kerans, takes up residence in the penthouse of the abandoned Ritz hotel, dressing in the silk shirts of the suite's previous resident, a Milanese financier, and helping himself to the cocktail bar.

The Drowned World paints a hallucinatory portrait of a London submerged by rising sea levels, the ice caps long melted, and overgrown with tropical vegetation; where tides of silt drift against the flooded buildings and albino alligators lurk in murky lagoons. It is, in Ballard's vision, a return to a previous geological epoch—the steamy swamps of the Triassic—and over the course of the book, Kerans feels himself pulled by forces beyond his comprehension, as something resurfaces from deep within his psyche. As all humanity rushes north to the relative cool of the poles, Kerans turns his gaze to the south, and moves off through the jungle, toward the burning sun. He scratches onto the side of an abandoned building: ALL IS WELL. He knows he won't survive for long.

On the balcony, I turn from the pool and look out instead across the buried city below. And as I do, I feel something shift inside me. Something like revelation, a premonition.

❧ 12 ❧

THE DELUGE AND
THE DESERT

Salton Sea, California, United States

S
undown in the desert. I walk out under a stained-glass sky.
Carmine, indigo, amber, and a pale, sweet green move
through the sky in soft and overlapping bands, sinking to
the ground as if spent, slipping behind the mountains with the last
of the day's light.

I've been traveling a long time. I feel distant, light-headed, my
faltering progress increasingly taking on the dizzying significance
of a dream, in all its heavy-handed imagery: driving fractured
roads past vacant lots, boarded houses, the basins of marinas emptied of water, and—having pulled up and left my rental car askew
across the road—I stagger down between the arms of two twin
piers that loom impotently over a dust-dry landscape. Somewhere
out there, I know, are the silvered remains of a sea, a sea in the

process of simmering away into nothing, leaving only a pale shadow in its place.

The silt here wears a hard rime of salt that gives way as I lower my weight onto it, like sun-crusted snow. All along the foreshore, such that it is, a filthy spume has dried in rippling tide lines that recede into the distance. I keep walking, moving at right angles to the faded shores, past corroded pillars encrusted with dirty white sediment that jut from the silt, picking my way over the battered, sun-bleached detritus that washed up long ago as tidewrack.

As I get farther out, my feet sink deeper into the thin, gray sand. When I look closer, I see it is not sand at all, but the dry bones of fish, pounded into shards, and the tiny, skull-like husks of barnacles. This is a foul place. The air is thick with brine and guano and decomposition. Even now, in the violet dusk, the heat is oppressive. But as I cross the crystallized flats, the water gleams into view, an impossible sea in the middle of the desert.

It is a poison lake whispering sweet nothings. It promises cool succor, quenched thirst. Despite what I know of this shimmering mirage—despite the stink and the rot and the waste that surrounds it, despite the staring eyes of the dead and desiccating fish that litter its shrinking shores, despite the absence of vegetation—I can't help but quicken my pace. I stumble through sucking mud toward this false vision, on and on until the muck is over my feet, and up to my ankles, and I am shin-deep in a warm broth that, when stirred, releases a draft so stagnant I can taste it.

I stagger back, holding a hand to my nose and mouth. Ahead, the dry hills of the opposite coast rise, arid and sculptural, as a ribbon along the horizon, all that separates vast prismatic sky from looking-glass sea. I look down, through the surface, and feel myself falling through the whole sky, all of it lit with a truly celestial light.

The Salton Sea is not a true sea, but the vestige of a great flood: the consequence of the Colorado River breaching the banks of an ill-built irrigation channel in 1905 and flowing down as a torrent into what was then the Salton Sink, a vast parched playa in southeast California that then stood bare and unadorned but for a few steaming, sulfurous springs.

The Colorado floodwaters were a near unstoppable force that carved deep gorges into the loose desert soils, and a waterfall thirty feet high that eroded its way backward through the basin floor at the speed of a man walking. Water overcame the small woodshed town of Salton, the salt pans that formed the valley's main industry, a railway siding, and eleven thousand acres of the Torres-Martinez Desert Cahuilla Indian tribe's ancestral lands. The waters rose and rose, as much as six inches a day, filling the valley like a bath and creating an inland sea thirty-five miles long and fifteen miles wide.

The residents turned out in hundreds to watch the deluge as it swallowed first their fields and then their homes. Though dramatic, the arrival of water in a region once known as "the Valley of the Dead" was not entirely unwelcome; "while naturally such an unexpected turn has caused a great deal of inconvenience," commented a local paper, "the flood will really prove a great benefit." The barren land might be irrigated, and some respite found from the desiccated air. "The effect of breezes blowing over the hot valley from a salt sea* would temper the climate to a delightful balminess that would make this part of the state one of the most attractive home spots to be found anywhere."

And so it transpired. By the 1950s, the accidental sea had bloomed into a popular resort, rechristened the "Salton Riviera." An hour's drive from the upscale nightclubs and golf courses of Palm Springs, the Salton Sea offered a yacht club, motels, water skiing, and—after the waters were stocked with shad, orange-mouth corvina, and striped mullet—sport fishing. For a time it saw more visitors annually than even Yosemite National Park. Naked children frolicked in its serene waters under cloudless skies. Enormous flocks of waterfowl adopted it as a stop-off in their seasonal migrations, congregating on the water in vast rafts: pelicans, grebes, avocets, plovers, ruddy ducks, even flamingos. So many birds came in during the migrations, I'm told, the sky would turn to black. You could track the time of year by the population of birds on the water; in spring, the white pelicans would gather, and swirl up off the water as one, sparkling like diamonds. It was, as one ad hailed it, "truly a miracle in the desert."

But the miracle was short lived. When it wasn't flash flooding—

---

*Though the waters of the Colorado River were fresh, a great deal of salt had lain locked up in the salt flats that lined the Salton Sink, and dissolved into the water on its arrival.

the tiny seaside resort of Bombay Beach flooded so badly and so often that the streets closest to the shore were eventually abandoned to their fate and a sea wall ten feet high was thrown up around the remainder of the town—the sea was boiling away into nothing. It began to shrink, revealing an expanse of heavy, clay-like sediment, which itself dried to a thin, alkaline powder laced with selenium and arsenic and DDT from the agricultural runoff that had been diverted into the sink to slow the sea's evaporation. Soon—baked dry and scoured by desert winds—the whole region became a dust-bowl, the toxic particulates soon triggering an asthma crisis across southeastern California.

It's a grim state of affairs that echoes that of the Aral Sea, the massive endorheic lake between Kazakhstan and Uzbekistan, which is commonly cited as the worst man-made environmental disaster in history. When a massive Soviet irrigation project diverted the flow of the Aral's tributaries into cotton fields in the 1960s, water levels immediately began to nosedive. Within a few decades fishing boats were left landwrecked in their ports, miles from shore; glutinous mudflats the color of mushroom soup dried into cracked plains; dust storms whipped up, thick with toxic pollutants—the residues of nuclear testing, heavy industry, and intensive agriculture. Now, the sea covers only 10 percent of its former area, and has been simmered down to a concentrated stew. Those who remain in the devastated region suffer from tuberculosis and various cancers at anomalous rates. Though recent efforts and a new dam in the northern reaches of the old sea have brought hope of a partial recovery there, it is at the expense of the

Uzbek south. "At first you drink water," as an old Uzbek saying goes, "at the end, poison."

As at the Aral Sea, the Salton waters have grown increasingly saline as the waters receded (at 5 percent salinity, it is now significantly saltier than the ocean) and tainted with both sewage and the byproducts of agriculture. In the 1980s, the accumulation of fertilizer in the salt lake began to have a curious effect: the nutrient-rich, alkaline waters grew lush with life—artificially so—supporting huge quantities of plankton and an estimated hundred million fish. For a while, this souped-up ecosystem was the most productive fishery in California; a single angler might hope to catch two hundred pounds of fish in an hour.

But under the heat of a desert sun, this roiling cauldron of nitrates and phosphates could all too easily boil over. Algal blooms exploded into life as cloudbursts of jade and malachite, billowing thick and vivid as paint. Or else the water might be muddied with strange hues, as if you'd rinsed your brushes in it—red like wine, like a good burgundy, or sometimes purple, even pink.

Every salt solution has its own bacterial familiars. When the waters drew back they revealed tilapia nests—dug by the fish like a honeycomb into the silt—each perfect round pool evaporating away at its own particular rate, taking on a different shade, color changing with the shifting concentrations of salt, an artist's palette that stretched for miles down the coast. Once, the Salton Sea was seen to glow: an aurora of the water in swirling cyan. Shortly after, huge rafts of white, bioluminescent foam washed up on the shore like a hallucinogenic bubble bath.

Such phenomena, however enchanting, are bad news for other sea life. As the blooms die off, they exhaust the oxygen in the water, leaving its fish gasping for breath and, ultimately, dying in enormous numbers. In the summer of 1999, ten million dead fish washed up on the Salton shore; three million in 2006. Their bloated bodies crowded the shallows, rotting, bleaching in the sun. Their flesh and fats broke down in the waters, decomposed, then washed up again as pebbles of adipocere—"corpse wax"—the size of satsumas.

These annual die-offs, and the hypersalinity of the water, caused the booming Salton fishery to collapse. By the time I visited, even the hardy tilapia were locked into terminal decline. And as the fish died, the birds that fed on them disappeared too, or worse—died off themselves. Avian botulism wiped out 10,000 pelicans in 1996, including 1,600 endangered brown pelicans. The following year, 2,000 double-crested cormorants—nearly every hatchling born that season—were wiped out by Newcastle disease. This was a sick ecosystem heaping sickness upon sickness in a cycle of destruction, an environment spinning dangerously out of control.

Algal blooms themselves can also produce potent toxins, including nerve agents. At the Salton Sea, these red tides have been linked to the deaths of more than 200,000 grebes. Harmful algal blooms have become much more frequent since the 1980s, both at the Salton Sea and, more generally, across the globe. The causes are disputed, but many point to the warming waters due to climate change and the increased use of fertilizers.

More common, and the source of the sea's dreadful odor, are the green tides: high winds stir up the anoxic waters at the lake bed—full of all those decomposing fish and rotting mats of algae—turning the water a gaudy green, suffocating all that touches it, and releasing large quantities of hydrogen sulfide—a lethal toxin with the stench of rotten eggs.

Altogether, these symptoms of environmental collapse add up to the atmospheric setting of a post-apocalyptic graphic novel: the toxic dust; the swirling, pigmented sea; the neurotoxic algae; the fish-bone beaches; the dissolving seafront trailers sinking into the mud; the jetties launching out into nothing. Except it's real, it's here, and it's only getting worse.

If there is an apocalypse, it will almost certainly smell like hydrogen sulfide. In the history of the Earth, there have been a number of what we call "mass-extinction events"—the best known being the Cretaceous-Palaeogene extinction, the meteorite impact that ended the reign of the dinosaurs. But not all such events rained down unannounced from the skies, *diaboli ex machina*, if you will.

The American paleontologist Peter Ward has written extensively on past extinction events with terrestrial causes—poisonous atmospheric states that have arisen from runaway feedback processes during global warmings. During "the big one," the end-Permian extinction—when 99 percent of all life forms were wiped out—ice caps melted, sea levels rose, and ocean currents ground to a halt. The deep sea warmed, became anoxic (as the much shallower bottom of the Salton Sea does in summer, when organic matter decomposes faster), and then this deoxygenated water slowly drifted up through the sea column, killing almost everything in the ocean. Normal bacteria cannot survive in anoxic water; instead, it supports a totally different bacterial flora—sulfur-utilizers—which proliferated and formed ocean-wide blooms belching poison gas.

"Hydrogen sulfide kills animals even at low concentrations," writes Ward, "and the rock record shows repeated episodes when large volumes of this gas came out of solution from the sea." Free-floating in the heated atmosphere, it killed most Permian land life "gruesomely," including plants. Such an event has happened at least eight times in the Earth's history, he stresses. Hydrogen sulfide release "may have been responsible for the majority of mass extinctions, and it certainly could happen again."

Ward views runaway processes like this—in which Earth systems, thrown out of kilter, form feedback loops that spiral rapidly out of control—as evidence of our planet's inherent suicidal tendencies. With a knowing nod to Lovelock and Margulis's comforting Gaia principle (Gaia, the goddess, being mother of all creation), Ward named his own hypothesis of algal immolation after Medea, the spurned barbarian who kills her own children in revenge for ill treatment.

Medea, says Ward, is the most appropriate representation of the planet's past behavior, which "has shown itself through many past episodes in deep time to the recent past, as well as in current behavior, to be inherently selfish and ultimately biocidal." Episodes like those seen at the Salton Sea—red tides killing a million fish, whose decomposition produces anoxic waters, hydrogen sulfide, and ultimately a green tide, killing even more fish—should be, under Ward's thinking, characterized as "Medean phenomena."

Humans, of course, busy sloshing fertilizers into our waterways, and—worse—pumping the atmosphere so full of carbon dioxide as to risk the toppling of the entire planetary system, may yet prove the most Medean species of all. To trigger sulfide-bacterial blooms on a planetary scale, Ward estimates carbon dioxide levels would need to reach 1,000 parts per million (ppm). Before the industrial age, levels of carbon dioxide in the atmosphere sat at around 280 ppm. By 2019, that figure had risen to 411 ppm. Under the most optimistic forecast put forward by the Intergovernmental Panel on Climate Change, we might hope to drag concentration levels back under 400 ppm sometime in the next century. Under the most pessimistic, these numbers could easily overshoot Ward's red line, spiking to a frankly horrifying 2,000 ppm by 2250.

Here, on the shores of the dying sea, where fish swim drunkenly in anoxic waters, gasping for breath, where the bones of their dead crumble into dust underfoot, such knowledge washes over one in a wave of regret and recrimination. It is hard not to see this place, this Medean landscape, as a portent: a vision of the future or warning of what more is to come. An augury of the end of the world, the dawning of the age of dust.

Maybe it was the plague-like proportions of the fish and bird die-offs, or the stench of rotting corpses. Maybe it was the noxious dust blowing through the streets. But by the 1990s, the settlements along the Salton Sea coasts had largely emptied out.

In Salton City, on the sea's west shore, a street plan is laid out ready for a prosperous future that never arrived. It's a bare skeleton of a town, where more than two hundred miles of roads, cracked and dry, slice uselessly through the empty land, crisscrossed with power lines, leading only to the occasional cinderblock house or trailer. Many of the buildings there are boarded or burned out, but at its heart sits an all-American high school—newly built and sparkling clean, with a football field out back, floodlit and luminous against the bald pate of the desert.

LAND SALE! shouts a peeling sign at the side of the road. No kidding, I think—but at some stage they must have been confident, because the footprint is enormous. I twist left then right, following empty streets with aspirational names down to the Riviera Keys, where houses back onto a drained marina and wooden jetties jut out into empty space. Boats lie upturned by the roadside. I drive on. Rumpled, buckling roads in curving suburban layouts lead me past houses, mostly empty, some neat and tidy but more often tumbledown. Some are occupied, revealed by the barking of a dog or the flicker of the television from a darkened room. Farther out, where the streets are no longer electrified, several RVs lie beached: one squat caravan tips as if riding an ocean wave; another's burned carcass casts a shadow of ash.

The road turns to track, turns to sand. I park up next to where an old sofa and armchair have been dumped, half-buried by a dune. The ground between the mesquite and the saltbush is bare and beige. Not far away, a jackrabbit lifts itself proudly onto its haunches, back arched, and regards me calmly before bounding away. There is rubbish everywhere—faded and beribboned by the heat.

A trembling percussion I can't place fills the air. I turn and meet with a bizarre sight: flies—hundreds of them, thousands of them—are raining down upon my car hood and roof. They pour down the bodywork, trickle down the windshield, drip into writhing puddles on the ground. I can't imagine where they are coming from; all I know is that they are, inarguably, there, as if summoned from the heavens. I step forward, into the spray, and feel a drumming as they rain down into my hair and onto my shoulders, wriggling as they slide down my neck and into my shirt. Appalled, I jump backward and find the car is trapped in a climate all its own: a plague of flies has been cast upon it, which does not follow as I move away. I watch, both fascinated and repulsed in equal measure, until finally I turn from the vehicle and hike up the slope.

I find myself on a scalloped, artificial ridge, a repeating landform that marks the edge of the old atomic test station and bombing range. This is where prototypes of the H-bomb dropped on Hiroshima were tested. Now long abandoned, it forms a restless expanse of drifting dunes and derelict bunkers. Two miles out to sea stands a metal platform once used as a target for their dummy runs, a relic of the atomic age. Impossible not to think of Ozymandias, the toppled warlord, in his desert tomb:

> Look on my Works, ye Mighty and despair!
> Nothing beside remains. Round the decay
> Of that colossal Wreck, boundless and bare
> The lone and level sands stretch far away.

Ahead of me, boundless and bare, stretches the dry lake bed coughing its toxins into the air. An environment wrecked. A moment passes, then another. Time elapses. After a while, I turn and go back the way I came, past the threadbare couches and piles of silvered beer cans, and into the hail of flies.

Seven miles east of the Salton Sea lies another abandoned military base dating from the Second World War. Camp Dunlop was built as a training base for an artillery regiment of the U.S. Marines; it featured eight miles of paved roads, a swimming pool, water tanks, and around thirty buildings. The forces vacated the site at the end of the war and took the buildings with them, leaving only the foundations they sat on. Now they call it Slab City.

Since the 1960s, this place has served as the site of a makeshift desert camp of dropouts and drifters, hippies, tweakers, artists, outlaws, runaways, survivalists—a haven or a hideout for those who have no home, or carry their home on their back, or have burned their homes down. It's busy here in winter, when snowbird pensioners in expensive rigs tumble through, a thousand at a time, looking for a free place to park. But when the heat amps up in summer, as high as 120°F, with no access to running water or power or sewers, they pack up and drive off. I get there in September at the end of a long, relentlessly hot summer, when only the residue, the hardcore, the true faithful, remain.

More shantytown than city, Slab City sits somewhere on a spectrum between commune and slum. Its grid layout, a hangover from its military origins, serves as the lone organizing principle for a hodgepodge of sun-bleached RVs, foil pasted across their windshields; shacks built of pallets and planks and tarpaulin; compounds built of corrugated sheets, rusted oil drums, and hard-worn tires; and water tanks hoisted onto wobbling stilts. Burned-out car chassis litter the road edges, shrouded by tattered saltbush, whose own branches are swaddled in plastic, shredded by the desert winds, fluttering like prayer flags.

It's squalid and ugly, but there's a raw splendor to the place too. Here and there, the heaps of refuse have been fashioned into works of art. There's a maze constructed of stones, piled into thin spiraling paths; a stripped-down car, propped up on bricks and with its hood gaping, ornamented in a thousand bottle caps like a pearly king; there are walls of glass bottles stacked lengthwise and cemented with clay, necks bristling like the fur of a frightened animal. Broken shards of mirror have been mosaicked back together to form a refractive, disconcerting whole.

Its residents call it, fondly, "the last free place in America." But it doesn't feel like a hangover from some untroubled past. If the slabs the squats are built on are souvenirs of the atomic age, then Slab City itself seems a vision of a post-atomic future: a hardscrabble society cobbled together from the ruins of a fallen civilization.

I stay with a guy I found online. Let's call him Jim. He's been minding the Slab City "hostel" over the summer in return for a place to live. It must be more appealing in winter, but when I arrive the hostel comprises a few dirty mattresses separated by hay bales and a stuffy, oven-hot Winnebago which Jim expressly recommends I avoid. I'm the only guest.

"Welcome to the Slabs," he says when I arrive. "Twelve people have died here this year. Also, they caught three murderers, two child molesters, and a couple of robbers hiding out."

I feel I'm being tested, raise my eyebrows, try to affect an expression that could signal either shock or skepticism. "Hah," he says, after a strange moment. "I'm just messing with you."

I make a movement with my mouth like, phew, to signal my relief.

"But there have been deaths," he quickly clarifies, as if he doesn't want me to get the wrong impression.

The deaths were from heatstroke or dehydration, mainly; unfortunates who set off on the four-mile hike to Niland, the nearest town—itself semi-derelict—and didn't make it. Or overdoses:

crystal meth, heroin. Or people pass out, drunk or stoned, and just never wake up. But Jim's joke about the ne'er-do-wells isn't wrong enough to be funny, either. A month before I arrived there, a Slab City regular was arrested on child sex charges—sodomy with a minor under ten, among other things—and five months before that, a fugitive wanted for the shooting of a nineteen-year-old in Virginia was caught hiding out here in a tent. But, says Jim, it's not like that. Not really.

Jim's in in his late forties, maybe early fifties, heavyset, wearing a tie-dyed shirt and an elasticated skirt—for the heat, he says, which is oppressive. Not long ago, he lost just about everything he owned in a fire. He gets a bit of money from the state thanks to his disability check, but it's not enough to live on anywhere else. He can't go home to Indiana: "I'm a bad man," he says, as if in explanation, and we leave it there. He's got a jar full of marijuana buds and two dogs, but no shoes and no car. His feet are as nut brown and gnarled as the roots of a tree.

Later, in a sculpture garden known as East Jesus, we see a rat-faced guy all in black lurking between the artworks and the mere sight of him is enough to make Jim jittery. He knows that guy, he says. He burned his camp down a few days ago cooking meth. He's got a machete and nothing to lose. Jim raps loudly on a nearby trailer and shouts until a girl comes out, looking pissed off. She nods once, grimly, when he explains about the lurker. It's cool, she says. I got company. And I got a shotgun.

All this is to say: it's an anarchic place. A refugee camp that will accept you whatever you are fleeing, be it personal demons, a bad marriage, or the law. The desert is a blank slate, as one man tells me. The Slabs are endlessly forgiving, no matter your crime; it simply doesn't ask.

To the south rises the Slabs' most famous landmark, Salvation Mountain—a hill-sized sculpture–cum–landform–cum–place of worship built of adobe and hay bales and painted in bright Sgt. Pepper colors by the late, great outsider artist Leonard Knight. Knight, a lifelong drifter, converted to Christianity in his thirties and spent three decades living out of his hand-ornamented truck at the foot of his growing psychedelic shrine in the middle of the desert.

GOD IS LOVE, it declares in huge bubble letters rolled from clay.

JESUS I'M A SINNER PLEASE COME UPON MY BODY AND INTO MY HEART.

REPENT, it instructs. REPENT NOW. A crucifix sprouts from its summit like a beanstalk.

As disconcerting as it is striking, Salvation Mountain is the work of a beautiful, unhinged mind: the work of an unlikely prophet who came to the wilderness, like so many before him, in search of God.

Salvation Mountain promises forgiveness, unconditional love, in the place that needs it most. Pilgrims arrive in Slab City seeking rebirth: a fresh start, a new life, a blank slate. A place to crash while they sort themselves out. They hope to shape something new out of the wreckage of the old, both physically and spiritually. People here go by aliases, whether they need to or not: "White

Horse," "Caveman," "Dreamcatcher." But though we all may repent, we may not always be forgiven. Sometimes the denizens of
Slab City learn the true names of their friends only when they
turn up in newsprint, when the past catches up with them after all.

I'd heard there were hot springs at the Slabs, but Jim warns me
off. They're *hot*, he points out, not unreasonably: at about 100°F,
they're the last thing you need when you're already light-headed
and sun-sick. Plus, it's not hygienic. They bubble away indefi-
nitely, a broth boiled off every Slab City resident who came before.
Or worse: a few years back, a dead body turned up in the murky
water—a popular young Slabber, whose corpse had lain submerged for hours. In that time, bathers came and went, soaked
next to it, unknowing.

So, I pass. But Jim says he has an alternative plan. We get in my
car and he directs me along a long gravel track behind the camp to
where the clear waters of the Coachella Canal flow fast through
a V-shaped concrete scar cut through the desert, water destined
for the swimming pools and sprinklered golf courses of Palm
Springs. I hesitate for a moment when I see it. It doesn't seem real.
All week I've been fending off the false flags of mirage as I travel
through the desert: shimmering visions of flooded roads that
retreat upon approach, and the floating islands of fata morgana.

But this is real. Really real. Cool, pure, fast-flowing water. I drop my grimy dress to the dirt and jump in near a ladder, which I use to anchor myself against the current. It wants to sweep me downstream and under a fence into some kind of mechanized weir. DANGER, reads a sign, but in my heightened state, the underlying risk seems only to intensify the experience: the water so clear, so turquoise, so temperate. Tiny fish shelter under the ladder rungs. Catfish sweep the smooth concrete bottom. The sky is so cloudless it appears black when I look up. I feel dizzy if I do, as if peering over the edge into a bottomless gorge.

I think of a woman I met earlier, Ella. She moved to the Slabs for health reasons: chronic pain and the OxyContin addiction that came after. But she felt reborn in the desert, in the hot, dry air. She said she'd never go home. I think I know how she feels. I am baptized, wiped clean. Jim takes a running jump, and strikes out in a stiff front crawl, angled into the current so that he cuts across to the ladder on the opposite wall. He grabs a rung, and hauls himself to safety. Heaving himself half out of the water, he throws his head back and crows like Peter Pan, a wild ululation. Then he laughs: he hasn't had a shower since July. No matter. The rich are drinking our bathwater now. This is how you can feel like you have everything when you have nothing at all.

Jim directs me a different route home. I'm trying to be helpful, to follow instructions obediently, but he seems suddenly high, or higher than before, and keeps jerking upright in the passenger seat as if from sleep. He seems suddenly disoriented although we're only a few hundred yards from where he's been living.

We end up, somehow, in the library—a shack-like structure built of repurposed wood and corrugated sheets, half-open to the elements. It's a beautiful concept—dreamed up and staffed by public-spirited Slabbers from their own resources—but nevertheless the space has a dusty, passed-over atmosphere: the books are stacked tightly, spines bleached, pages swollen. A thick felt of grime covers everything, like a lost library from centuries ago.

The selection is haphazard, a mix of titles mostly of the sort that might fill bins in bargain basements and charity shops, many of them long out of date, so it's odd to see them here, out of context, elevated by their inclusion upon the library shelves. It makes me think of the "museum" of everyday objects described in Emily St. John Mandel's *Station Eleven*, a novel in which survivors of a deadly global flu pandemic carve out a new life in an abandoned airport, and—over time, as items are rendered obsolete by the collapse of civilization—amass a collection of artifacts dating from before the fall: passports to forgotten countries, mobile phones without a signal to connect to, credit cards keeping score in economies long collapsed, stilettos, laptops, stamps.

Here, there are faded Catherine Cooksons, *Reader's Digest* anthologies, aging magazines curling at the edges—but in the dusty gloom, they have the same effect as St. John Mandel's museum: that of wistful nostalgia, of significance in objects, of the partialness of history pieced together from fragments. In one corner, there's a cylindrical stack of aged encyclopedias with a spray-paint label: GOOGLE.

If anyone has the skills to survive some unspecified global disaster, they are likely to be found among the residents of Slab City, who are living now as if the end times have already come. The overwhelming aesthetic—partly self-aware, but largely through necessity—is of a post-apocalyptic wasteland with a *Mad Max* vibe, where buildings and machines fallen into disrepair have been cannibalized and then cobbled back together. This is an off-grid desert camp, where dust storms scour every surface in summer, and flash floods blast down arroyos and turn ground into quicksand in winter. To get by you need shelter from the sun, a means to store water, and a means of protecting your property—weapon or guard dog—and not necessarily in that order.

Eschatology—and eschatological storytelling—has often flourished during times of crisis, and a lot, I think, can be gleaned about a culture by studying its particular brands of apocalypse. In recent years, popular culture in the West has been increasingly dominated by dystopian visions, both in cinema and in literature, including a rise in so-called "cli-fi"—fantastical visions of climatological disaster, which echo real-life anxieties over the impact of man upon the planet.

And it cuts both ways; it's impossible not to see parallels in the most vociferous of climate change literature: there too we find that sense of impending disaster, of divine retribution for past sins, the urgent need to act before it is too late. Professor Jem Bendell, of the University of Cumbria, has shot to prominence in certain ecological circles thanks to his darkly prophetic writings that assert that "climate-induced societal collapse is now inevitable in the near term." Bendell builds his doomsday case on a solid foundation of peer-reviewed papers, and paints a picture of a dystopian near-future in which "mass starvation, disease, flooding, storm destruction, forced migration and war" feature heavily.

What we call "civilization," he adds, "may also degrade." Social collapse will be the norm in most countries within a decade. It is too late to avert the disaster; we must make our peace, and save what we can.

Bendell asks, rhetorically: "Where and when will the collapse or catastrophe begin?" and I too cannot help but imagine a rapture sweeping the globe; starting perhaps from the low-lying land and washing inward; settlements collapsing in its wake, survivors clinging to the ruins, to the shallow facsimiles of contemporary culture. How would it unfold, I wonder: the creeping decline, or the sudden collapse? Would we one day find the shop shelves bare? (In the spring of 2020, we learned how fast that can happen.) How long until the phone lines go silent, until the taps run dry, until we lose the ability to access our hard drives and we display them on shelves built of planks in the Slab City library?

Bendell's initial paper, which proclaimed the need for "deep adaptation"—that is, acceptance of the collapse that is to come—was rejected for publication at a mainstream journal, but after he self-published it online, it went viral: downloaded hundreds of thousands of times, as word of his dark prophecy spread. The appetite for such a document underlines how deep, cloaked within the terror of the eschaton, lie perverse desires: the thrill of danger, and the reassurance that disaster might be averted should certain strictures be followed; or that a select few among the faithful might hope to survive the onslaught to build a better, truer life on the other side. It reflects, perhaps, a *yearning* for apocalypse, or certainly a growing expectation of one. One 2019 poll found that 51 percent of Americans under the age of thirty-five believed that "it is at least somewhat likely that the Earth will become uninhabitable and humanity will be wiped out" by climate change within the next ten to fifteen years. (A matter of nominative determinism, perhaps, for a generation already known as "millennials.")

The Stanford biologists Paul and Anne Ehrlich, who infamously predicted in *The Population Bomb* that hundreds of millions would starve to death in the 1970s, have redirected their anxiety around overpopulation into questions of climate. A consequence of the doubling of the human population has seen a tripling of consumption. The results, they say, indicate Malthusian catastrophe of a climatological nature is not far away.

On examining the graph of human population growth, the near exponential burst over recent decades, it's hard not to visualize in response all those boom and bust cycles seen among invasive species, and to brace for impact. But there is sometimes too a hint of glee to the way this prospect is discussed in certain environmental circles, among those who view humans as the greatest invasive species of them all.

An artist I know, a gifted woman with a brilliant mind, told me once over dinner that she didn't believe in giving children vaccinations. I didn't want to start a debate, but couldn't help but voice my surprise. "I'm an environmentalist," she said in response. "I think the world's in need of a good plague." I was taken aback by her directness, but also struck by the purity of her beliefs, her apparent willingness to follow them to their logical conclusion. Anti-vax as praxis. It gave me pause for thought. Though her take is unusually direct—in actively courting disaster, almost inviting it upon her own family—I have heard others offer similar arguments in more abstract terms. A crash in human population, the thinking goes, and the resultant let-up in demand for the Earth's resources, the decline in the burning of fossil fuels, and the massive regrowth of agricultural lands, will lead to the recovering of the planet.

During the COVID-19 pandemic, the grassroots campaign group Extinction Rebellion was forced to distance itself from what appeared to be a wayward regional chapter celebrating the deaths of coronavirus victims. ("Corona is the cure," read flyers pictured on social media. "Humans are the disease." The account was later deleted.)

The appeal of disanthropic thinking, I feel, lies in the notion that a crash in the number of humans might present an opportunity akin to pressing a reset button. The fantasy is this: you and I survive, and together we start again. Do it better this time. It's a seductive argument, and, in the case of the Black Death, this was not *un*true. (Medieval society was radically reconfigured in the decades that followed the horrors of the plague; the system of serfdom collapsed due to the lack of laborers, and the lowest classes gained higher wages and better access to land and resources.) But it is also a line of thought both hubristic and inhumane in its disregard for the hurricane of tragedy and grief that would unfold, and sweep all of us up into it, however well we might prepare.

"If nature dies because we enter it," the environmental historian William Cronon wrote in 1995, "then the only way to save nature is to kill ourselves." Cronon was not arguing the point, rather skewering the absurdity he saw at the heart of purist environmentalist thought—those who feel humanity itself to be sullied, pollution, the original sin. But this notion has been taken up, extended by others—not least those of the Voluntary Human Extinction Movement, who practice self-sterilization (motto: "May we live long and die out").

Rupert Birkin, the disanthropic love interest in D. H. Lawrence's *Women in Love*, gave voice to this conviction: "Well, if mankind is destroyed, if our race is destroyed like Sodom, and there is this beautiful evening with the luminous land and trees, I am satisfied. . . . Let mankind pass away—time it did."

Not hard to picture here, where the Salton stink fills the air when the wind blows in a certain direction. Hard too not to think of the fire and the brimstone and those biblical kings fleeing their ruined cities over the sulfurous mudflats of the Dead Sea.

In the library, three topless men sit at a table, shooting the breeze. They nod a greeting to Jim, and ignore me. The man speaking wears his hair pulled back in a ponytail, but shorn at the sides. "The other night I got up for a piss," he's telling the others, "and I almost didn't take a flashlight. Thank fuck that I did." He leans over, and, as he does, the shadowed space is filled suddenly with a susurrus I can't place but floods me instantly with anxiety.

My eyes dart everywhere as I search, spooked, for the source of the sound. The men see me and laugh. The library's guardian produces what looks like a garden cane from under the table and shivers it in my face.

"Jesus," I say, "What *is* that?" I take the stick to examine it: the rattle from a rattlesnake's tail has been cut off and pinned to it. He quivers the cane again and it produces a noise like a rain stick.

"It was sitting waiting for me in the middle of the floor in the dark. So I shot it." He stands up and fetches a board where the snake's skin has been stretched out to cure.

What happened to the rest of it, I want to know. He ate it, he says. The second man, who wears a black crucifix around his neck on a leather thong, laughs a dry laugh that feels at my expense.

It tasted like meat, the librarian continues. Like chicken, or pork. But the flesh flaked off the bone like fish. He gets up, fetches something else to show me: an old plastic fish tank of the kind a child might own. I peer through the scratched panel to see the dry husks of two dead scorpions, locked in suicidal embrace. Shit, he says, when he realizes what we're looking at: I guess they were fighting.

The third man at the table is younger, fresh-faced, with a look of bland good humor. I don't pay him much attention until I bang into him later at the foot of Salvation Mountain, and find him manning a folding table under the shade of an awning. He seems pleased to see me again.

"2K," he says. His name. He's wearing a turquoise earring, a Palestinian scarf, and a skirt of checked material hitched up with a canvas belt. His hair, shorn on the sides, is golden brown, and almost exactly the same shade as his deep, outdoorsman's tan. 2K tells me he's from Missouri, and got here at the start of the summer after some time on the road. Before he left home, he'd been working—of all things—at a doggie daycare facility, but one day he realized he couldn't do it anymore.

"I was tired of the whole Babylon thing," he tells me. I shake my head. Babylon? Babylon, he says again. The outside world. People working all the time. Working to live, living to work. Working to make money to spend money. The whole capitalist cycle. So, he opted out. He quit. He packed everything he owned into his truck and set off.

But when his last paycheck came through, it was less than expected. The vehicle was unreliable. He got stranded in Indiana, waiting out a blizzard under a blanket with his dogs. In the end, he abandoned the truck in New Mexico and hitchhiked west. Two golfers on their way to Palm Springs picked him up. And so, in the end, he washed up here.

He slept under a bush a few nights before he got here, he says. Now he sleeps under a tree. "I've gone up in the world," he says, and it's a joke, but it's also for real. There isn't much shade under the tree, so he spends most days under a tarpaulin at a neighbor's camp or in the library. When the temperature drops, he'll build an adobe hut, from straw and clay, to live in next year.

Living rough like this, he sees a lot of desert life. Corn snakes, gopher snakes, shiny-skinned skinks, white desert iguanas, quail, jackrabbits, camel spiders, black widows, doves. Mainly, though, it's deer flies. They bite, draw blood. It sounds grueling to me, but he seems light on his feet. "There's so much available to me here," he says, with a flash of evangelical zeal. "I'm almost overwhelmed."

Other Slabbers, he says, are "expressive." You've got to keep safe, keep an open mind. Others still prefer to keep their own company. He waves vaguely across a dirt track to where a few lone shacks, set far apart, loom darkly in the distance: leaning towers constructed of pallets and flapping tarpaulins. Desert hermits, retreating ever further from the world.

"Here in Slab City we have a tremendous advantage," says 2K, eyeing their constructions. "All the trash from Babylon." I look at him quizzically. He's not being humorous. "So much trash accumulated in the desert over the years it became a resource. Everything we have here has been built out of trash."

Babylon, to 2K, is all modern civilization: the great polluting monolith from which they live downstream, picking through the waste spewing from its exhausts. It is a chaotic, careless place of inconstancy and disappointment which squeezes its workers dry and then deserts them. The people who end up here come partly from choice, in a quest for a different way of life, and partly because they have nowhere left to fall. Slab City may be an abandoned, trash-strewn hinterland, but at least the rent is free. Babylon is what the notice on the old sentry box on the road into town is referring to when it says, WARNING: REALITY AHEAD.

Babylon was a real place, a wealthy kingdom in Mesopotamia; but in the Bible it represents man's wickedness and conceit, his arrogance in presuming to be better than God. It is a golden cup from which all humanity has drunk, appearing in scripture first as the site of the Tower of Babel, and last in the Book of Revelation— written long after the real-life fall of Babylon at the hands of the invading Persians—in the form of the eponymous whore, bedecked with gold and drunk on the blood of saints, mounted on a scarlet beast. "How much she hath glorified herself, and lived deliciously," writes the prophet John:

> *Therefore shall her plagues come in one day,*
> *Death, and mourning, and famine;*
> *And she shall be utterly burned with fire*

And all the kings of the Earth shall smell the smoke of her burning, and bewail her loss,

> *saying, Alas, alas, that great city Babylon, that mighty city!*
> *For in one hour is thy judgment come.*

I leave 2K at Salvation Mountain, but what he said stays with me all the same: a sense of the encampment at the Slabs as an inevitable byproduct of a reckless and profligate society. But more than that: of Slab City as a place of renunciation, of conscientious objection, where one might lodge one's withdrawal of support for the modern way of life.

As I pass, my eyes track the ramshackle cells of the anchorites. Who knows what they are driven by—God or drugs or mental breakdown—but they are not alone in their practice of desert retreat. It was a biblical wilderness, a desolate, waterless, featureless desert, where Moses saw the burning bush, where Elijah wandered when he alone was left among the faithful, where the Lord came to him "in a still, small voice."

I no longer have a god. But I feel him sometimes, anyway.

I did know thee in the wilderness, in the land of great drought.

In the desert, fear must turn into faith.

Faith, in the end, is what environmentalism boils down to. Faith in the possibility of change, the prospect of a better future—for green shoots from the rubble, fresh water in the desert. And our faith is often tested. The doomsday theses of Jem Bendell, Paul Ehrlich, and their ilk are not born of ambition or spite, but drawn from observation, careful study. From the facts, in other words, as well as we can make them out.

But I cannot accept their conclusions. To do so is to abandon hope, to accept the inevitability of a fallen world, a ruinous future.

And yet everywhere I have looked, everywhere I have been—places bent and broken, despoiled and desolate, polluted and poisoned—I have found new life springing from the wreckage of the old, life all the stranger and more valuable for its resilience.

This is a corrupted world, yes—one long fallen from a state of grace—but it is a world too that knows how to live. It has a great capacity for repair, for recovery, for forgiveness—of a sort—if we can only learn to let it do so. Lands cleared for cultivation centuries before revert to forest in a matter of years. Environments stripped of their inhabitants can repopulate of their own accord. Even contaminated sites of the very worst kind can, when given the chance, become ecosystems of singular importance.

At the start of this book I told you a story about the infant exiles and their mute nursemaid on the island of Inchkeith. That trial was known as the "forbidden experiment" for the cruelty of inflicting such extreme and lifelong suffering upon its subjects. In a sense, many of the sites I discuss in this book are forbidden experiments too—though conducted unwittingly, they nevertheless offer insights into processes too immoral to otherwise induce: nuclear meltdown, toxic contamination, stalemate warfare, political and social collapse. And, as with the infants, the results are striking, though difficult to interpret.

When I have described my observations to friends, on return from my various journeys, some have challenged my focus upon the positives—on the remarkable ecological recoveries—and voiced doubts over whether in doing so I risk undermining the hard work of many tireless campaigners and lawmakers who have identified and prosecuted those who have damaged the environment in count-less irrevocable ways. I think it's important, therefore, to underline that in doing so I am not proffering a free pass to those who wish to further pillage our planet. To me, these narratives of redemption offer something different: they are torches burning in a darkened landscape, beacons of hope in a world that sometimes feels bereft of it. They remind us of the power inherent in the world around us. And they are studies too of the benefits of sometimes surrendering control.

We have a tendency to roll up our sleeves, to "get involved" in natural processes—often on the basis that sites have been im-pacted by humans in the past: that it is "our responsibility" to undo the damage wreaked by our kind. But these sites remind us of the value of holding off from some of our most invasive, intervention-ist methods of conservation.

It's a difficult balance to strike, and the debate has its precursor in the field of medicine. In the nineteenth century, in response to the popularity of so-called heroic treatments—bleeding, purg-ing, and sweating cures—many physicians came to view medical intervention with increasing suspicion. These treatments, noted critics, were often harmful in themselves. Would it not be better, they argued, to abstain, and place our faith instead upon the *vis medicatrix naturae?** This manner of thinking, known as "therapeu-tic nihilism," proliferated for a time before the advent of medical testing provided empirical evidence of drug effectiveness. But

*Literally: the healing power of nature. The body's innate capacity for healing.

both schools of thought have a certain basis in truth. The modern iteration of the Hippocratic oath acknowledges this, exhorting physicians to avoid the "twin traps of overtreatment and therapeutic nihilism."

Our collective guilt over man's impact upon the environment can, I feel, propel us in the direction of overtreatment, based on an assumption that we know what's best for damaged habitats, and that it's better to do something than nothing at all. But the surprising vitality of abandoned sites—even those that appear, to the untrained eye, derelict and dull—and the way some can come to surpass carefully tended reserves in terms of biodiversity, demonstrate how interventions—as bleeding and purging did before them—can sometimes cause more harm than good. We must learn restraint, in other words: to recognize when best to give the Earth its head, as we might a horse in rough terrain.

The great biologist E. O. Wilson proposes that we might surrender half the Earth's surface to nature as a bulwark against future disaster—as a storehouse of biodiversity. In this, he draws from his theory of island biogeography, which holds that the greater the area, the wider the array of species a piece of land might support. He too thinks of "islands" in the metaphorical sense, and his breakthrough findings have been of great inspiration to the contemporary rewilding movement, which now sets its goals at the "landscape scale."

But the islands of abandonment in this book serve to remind us that it is not only big, structured conservation projects that offer a return of the wild, but the scrappy abandoned parking lot at the end of your road. Consider it and every one like it a tiny islet in an archipelago stretching over the whole world. Stepping stones for species as they recolonize what land was lost.

Places like Bikini Atoll and Chernobyl and the Lothian bings show us that the absence of man is often all the stimulus required to start the resurrection. Time is, after all, the great healer. The question is: *How long does it need?* Then: *How long have we got?*

It may not be long. This is the time of the confessional, of the admission of sins. This is a time to pray, if you know how: to God or to Gaia, or simply by throwing one's entreaties to the ether in a gesture of helplessness and hope. A time for faith, in other words.

Julian of Norwich, in the account of her thirteenth revelation, said that God came to her in corporeal form and told her that there must be sin in the world—it "behooved that there should be sin"—and yet, despite it all, everything was going to be all right:

> *that all shall be well, and all shall be well, and all manner of thing shall be well.*

I am no mystic. I have received no visitation, no annunciation. There may be no absolution. But I do know this: all is not lost.

At night, I lie awake in my hammock at the hostel. It's too hot to sleep, but overhead the desert sky is violet and pierced through with light. Hours pass in a haze of sweat and starlight and discomfort.

What consumes my thoughts is a throwaway comment from an article I once read, in which an artist described the Salton Basin as passing through "a death, rebirth cycle." I had batted it away at the time as the sort of meaningless blurb one often finds in artistic mission statements, but the notion has transpired to have a certain basis in truth.

Over thousands of years, the Salton Basin has repeatedly flooded and evaporated as the Colorado River silts up and diverts temporarily from its route to the sea. In its earlier iterations, the waterbody has been so huge as to encompass what is now the city of Mexicali, across the border in Mexico. This latest, man-made deluge, therefore, is only the latest in a cycle that has spooled out over millennia, and the stinking, shrinking Salton Sea only the latest iteration of the ancient Lake Cahuilla, whose phantom shorelines still line the hills like rings around a bath. Though aggravated by the nutrients in the agricultural runoff, the evaporation of the sea and the white-caked precipitate lining the sink is part of a natural process, as the sea sinks exhausted into the earth, as its predecessors did before it.

As the shad and mullet died off, and now with tilapia on their way out too, all that will survive in the waters of the Salton Sea will be the tiny, iridescent desert pupfish. An endangered species, the desert pupfish has remarkable capacities to get by in extreme conditions. It can survive in water heated to 107°F or higher, in salinity more than twice that of the ocean, and in low-oxygen waters such as that associated with algal blooms. The ancestors of these pupfish once swam in the fresh water of Lake Cahuilla, but adapted to saline and hung on in the hot springs and salt streams left when it evaporated away—an evolutionary change some have likened to humans adapting to drink gasoline. These superpowers of adaptation are what's allowed them to cling on, and wait for conditions to improve.

After the valley reflooded in 1905, the pupfish leaped into action. By 1950, fishermen were reporting shoals ten thousand strong. And even now, with the sea in terminal decline, the pupfish retain the potential to rebound at a moment's notice. In 2006, geological scientists studying experimental ponds in the Salton Sea—which had been carefully screened to exclude fish—found that within months the desert pupfish had infiltrated and were breeding prolifically. By 2010, pupfish in the ponds numbered over a million.

The pupfish's success recalls those technicolor tilapia nests: the changing colors of the water reflecting the changing salinity inside: each tone revealing the presence of a different type of halophilic microbe, blooming as the salt reaches an optimum concentration, then dying away. They remind us not to write off what appears "barren" or "uninhabitable." Habitability is a matter of taste. Creatures like the desert pupfish are marginal now. But, should conditions change, these are the species that might stand to benefit—even come into ascendance.

I see the flooding and drying, flooding and drying of the Salton Sink as a metaphor for deep time, the passing of epochs. Our planet, with its warming climate, could be locked now into a death phase, with mass extinction ahead, leaving behind only the nimble, the fleet-footed, the fast-to-adapt. In time, the planet too may ascend again into life. Every major extinction event on the planet has been succeeded by a burst of evolutionary creativity: rapid diversification as heretofore insignificant species take on the roles left empty by those wiped out by meteor or climate change or supervolcano. Should half the world's species be wiped out, new creatures will grow up in their place. But it could take a million years or more. We—as individuals, certainly, but perhaps as a species—may not be around to see it.

Climate disaster, then, could spell a long uncomfortable descent into a death phase that to us will feel unending. To risk it is reckless, a willful act of self-harm. And yet, this reluctance to act. This reluctance stems, I think, from something like denial—even among those who believe the science; an unshakable belief in its impossibility.

But then, think of this: the citizens of the ancient city of Babylon once believed their city walls to be impenetrable. There were two walls, one layered in behind the other, each twenty-two feet thick, the whole thing encircled by a deep moat. Xenophon tells us that Cyrus the Great, leader of the Persian army, arrived and laid siege to the city. The Babylonians, knowing they had supplies enough for twenty years, paid him no heed. One night, while all inside were celebrating a great feast, Cyrus ordered his troops to redirect the Euphrates river, which flowed through the city, into their trenches, and thus marched in under the walls on the dry bed of the river. While the Babylonians were celebrating, their enemies were rising up within their own great city.

While I've been writing this book, fires have burned huge

swaths of the Amazon, the "lungs of the planet," and millions of acres of Australia. In Australia alone, those fires released 400 million tons of carbon, more than the country's annual emissions. They too released vast amounts of soot and carbon dioxide, reducing the planet's reflectivity and sealing in heat. In the United States, the National Oceanic and Atmospheric Administration reported that the world's permafrost is now thought to be emitting between 300 million and 600 million tons of carbon dioxide every year as it thaws. Such factors are what we must fear most: feedback loops that lock in the climate's descent.

It may already have begun; the enemy may already be inside the walls. But we must find faith enough to fight. *For in one hour is thy judgment come.*

ACKNOWLEDGMENTS

A great many people assisted me during the research and writing of this book. So many, in fact, that I fear I will leave out some crucial acknowledgment. (If that is you: I'm sorry—and thank you.)

Thanks are definitely due to: Ron Morris; to Iain Hamlin, Bruce Philp, and Roger Hissett; for Dan Crowe at Avaunt (and Robert Macfarlane, who introduced us), whose commission to write about the Slate Islands first got me thinking about *jolie laide* landscapes and land art; to Yiannakis Rousos (also known as Mr. John, if you would like to book him as a guide); to Dr. Salih Gücel of the Near East University; to Paul Dobraszczyk; to Tarmo Pilving; to Ludmilla Juraschko (send my love to the dogs); to Ivan Ivanovitch; to Elise, Dave, and everyone at the Michigan Humane Society, who take such good care of their animals and made me feel so welcome; to Tom Nardone, petrol head and local hero—whose new project, Trash Fishing, aims to clear Detroit's waterways of junk, and deserves your support; to Constance M. King; and to the excellent Detroit-based journalist Ellen Piligian, who conducted this interview on my behalf; to Jesse Welter; to Wheeler Antabanez, who so kindly gave up his time to show me around the ruined mills of Paterson, New Jersey; to Ari Nath and Marissa Piro, for putting me up in New York; to Angharad Selway, Hamish Osborn, and Daniel Guest at Natural Resources Wales; to Alloyce Mkongewa, my excellent guide in the Amani reserve (and who is a very capable field assistant, should that be of interest); to Robert Mayombo and all the staff at the Amani Forest Reserve guesthouse; to Martin Kimwere; to Pierre Binggeli and Charles Kilawe, for their company in the cloud forest and their expertise; to William Annal (see www.swona.net for more details of the Swona Heritage Trust); to Hamish Mowatt; to Stephen Hall, expert on wild cattle; to Clover, David, Sunny, and Kristina; to Rod Stewart at the Montserrat Volcano Observatory; to the two Montserratian policemen who picked up a very stressed hitchhiker and got her back to the airport on time; to "Jim," 2K, Ella, Hornfeather, and all those at Slab City who spoke to me about their way of life.

It is always a nerve-racking prospect to wade into a very technical, or contentious, subject with the intention of carving out a clear narrative for a mainstream audience. So I must thank two academics—Professor Glenn Matlack at the Ohio Center for Ecology and Evolutionary Studies, and Professor Jim Smith at the University of Plymouth's School of Geography, Earth and Environmental Sciences—for agreeing to read over my chapters on forest succession and the Chernobyl exclusion zone with a critical eye. Any remaining errors are, inevitably, my own.

Acknowledgments

This, my second book, had a long and meandering genesis. I must thank several organizations that assisted me financially during that period: Creative Scotland (whose support has twice now given me the confidence and impetus to put pen to paper); the Society of Authors, who awarded me the John Heygate Award for travel writing in 2017, reinvigorating me when I was at a low ebb; the Calderwood Charitable Foundation (via their "Art of Journalism" initiative); the Whiting Foundation; and the Traulsen International Travel Fund. I was lucky enough to work on this book while in residence at Gladstone's Library in Wales; at the MacDowell Colony in New Hampshire; and at the Jan Michalski Foundation in Switzerland; all of these are inspiring centers for creative thought, and have made a deep impact on me. I encourage other writers to apply.

The very first iteration of the idea for this book was written and sent from East Rhidorroch in the wild west highlands of Scotland—thank you to our friends and consummate hosts Iona Scobie and Julien Legrand. Louise Gray and Will McCallum—friends, fellow writers, and tireless environmentalists—were both early readers, and have been a great source of support and enthusiasm. To Alex Christofi and Samira Shackle: imagine if we'd known in Freshers' Week in 2005 that by 2020 we would all have books published within a fortnight of each other. It's all I could have hoped for and more, and I'm so glad we've done it together.

To my agents Sophie Lambert and Amelia Atlas: I can't thank you enough for your support and clear-eyed advice. And to everyone at Viking, especially Emily Wunderlich, Nidhi Pugalia, Alicia Cooper, Christina Caruccio, Paul Buckley, Maya Baran, Nora Alice Demick, and Mary Stone—thank you. I'm so proud to be a Viking author.

To my family: my parents Fiona and Derek Flyn, my brothers Martyn and Rory, their partners Mel and Claire, and their children Maisie, Daniel, and Lucy. Thank you for everything. And special thanks to Richard West, who has accompanied me throughout this entire process, counseled me through crises, held my hand, and generally tolerated my increasingly antisocial behavior during my long dark night of the soul as deadline approached. You are patient, insightful, clever, and kind, and I couldn't do any of it without you.

NOTES

Invocation

2　"school of the prophets": Hamish Haswell-Smith, *The Scottish Islands* (1996; repr., Edinburgh: Canongate, 2015), 503.

2　"till God provided for their health": Hugo Arnot, *The History of Edinburgh, from the Earliest Accounts to the Present Time* (Edinburgh: William Creech, 1788), 260.

3　"but I knaw not": Robert Lindsay, in *The Historie and Cronicles of Scotland*, vol. 1 (Edinburgh: William Blackwood and Sons, collected 1899), lxix, states that the observations contained in this book come from several manuscripts of "unequal date" and probably of two different hands. Elsewhere given as "Robert Lindesay of Pitscottie" and reported thus: "Sum ayis they spak goode hebrew bot as to my self I knaw not bot be the authoris reherse," quoted in R. N. Campbell and R. Grieve, "Royal Investigations of the Origin of Language," *Historiographia Linguistica* 9, no. 1–2 (1982): 43–74, 241, doi:10.1075/hl.9:1-2:04cam.

3　a "brutish babble": John Pinkerton, *The History of Scotland from the Accession of the House of Stuart to that of Mary* (London: C. Dilly, 1797), cited in Campbell and Grieve, "Royal Investigations of the Origin of Language."

6　Antarctic ice and deep-sea sediments: See "Microplastics and persistent fluorinated chemicals in the Antarctic," Greenpeace, 2018; A. Kelly, D. Lannuzel, T. Rodemann, K. M. Meiners, and H. J. Auman, "Microplastic contamination in east Antarctic sea ice," *Marine Pollution Bulletin* 154, no. 111130 (2020), doi:10.1016/j.marpolbul.2020:111130; also C. L. Waller, H. J. Griffiths, C. M. Waluda, S. E. Thorpe, I. Loaiza, B. Moreno et al., "Microplastics in the Antarctic marine system: an emerging area of research," *Science of the Total Environment* 598 (2017): 220–27, doi:10.1016/j.scitotenv.2017:03.283.

6　properties are already abandoned: Government statistics quoted in Mari Shibata, "What will Japan do with all of its empty 'ghost' homes?" *BBC Work Life*, October 31, 2019.

6　all housing stock by 2033: Hidetaka Yoneyama, "Vacant Housing Rate Forecast and Effects of Vacant Homes Special Measures Act: Vacant Housing Rates of Tokyo and Japan in 20 Years," Fujitsu Research Group, June 30, 2015.

7   **by the end of this century:** F. Isbell, D. Tilman, P. B. Reich, and A. T. Clark, "Deficits of biodiversity and productivity linger a century after agricultural abandonment," *Nature Ecology & Evolution* 3, no. 11 (2019): 1533–38, doi:10.1038/s41559-019-1012-1.

7   **"mitigate a sixth mass extinction":** Isbell, Tilman, et al., "Deficits of biodiversity."

9   **ruin questing, a "heartless pastime":** Quoted in Rose Macauley, *Pleasure of Ruins* (New York: Barnes & Noble, 1953 [1996]), xvii.

9   **"spoken for the last time":** David Eagleman, "Metamorphosis," in *Sum: Forty Tales from the Afterlives* (Edinburgh: Canongate, 2009), 23.

10   **indicator species pointing into the past:** See Ian D. Rotherham, *Shadow Woods: A Search for Lost Landscapes* (Sheffield: Wildtrack Publishing, 2018).

PART ONE: IN ABSENTIA

Chapter 1: The Waste Land

15   **world's leading oil producer:** Barbra Harvie, "Historical review paper the shale-oil industry in Scotland 1858–1962, I: Geology and history," *Oil Shale* 28, no. 1 (2010): 78–84, doi: 10.3176/oil.2010.4.08.

15–16   **six hundred thousand barrels of oil a year:** Barbra Harvie, "The Mechanisms and Processes of Vegetation Dynamics on Oil-Shale Spoil Bings in West Lothian, Scotland," PhD Thesis, December 2004.

17   **lies entombed beneath the shale:** Kirsteen Miller, "The Disappearance of Westwood House," *West Lothian's Story*, December 11, 2018, available online at: https://westlothiansstory.home.blog/2018/12/11/the-disappearance-of-westwood-house/.

20   **"Dull roots with spring rain":** T. S. Eliot, "1: The Burial of the Dead," *The Waste Land and Other Poems* (London: Faber and Faber, 1940), lines 1–4.

20   **by agriculture and urban development:** Barbra Harvie, "West Lothian Biodiversity Action Plan: Oil Shale Bings," published on behalf of the West Lothian Local Biodiversity Action Plan Partnership, West Lothian Council, Linlithgow, Scotland, 2005.

20   **the tiny bing at Mid Breich:** Barbra Harvie, "The importance of the oil-shale Bings of West Lothian, Scotland, to local and national biodiversity," *Botanical Journal of Scotland* 58, no. 1 (2006), 35–47.

21   **"Out of this stony rubbish?":** Eliot, "1: The Burial of the Dead," lines 19–20.

21   **Otherworld, or the holy grail:** Leo Shapiro, "The Medievalism of T. S. Eliot," *Poetry* 56, no. 4 (1940): 202–13.

23   **"the green mantle of vegetation":** E. J. Salisbury, "The Flora of bombed areas," cited in "Long Live the Weeds," Kew Gardens, London, November 30, 2018, available online at: https://www.kew.org/read-and-watch/long-live-the-weeds.

Notes

24 **dropped on Hiroshima:** United Nations Educational, Scientific and Cultural Organization (UNESCO), "Bikini Atoll Nuclear Test Site," UNESCO World Heritage, https://whc.unesco.org/en/list/1339.

24 **a mile across and 260 feet deep:** Mike Carlowicz and David K. Lynch, "Revisiting Bikini Atoll," NASA Earth Observatory, March 1, 2014, https://earthobservatory.nasa.gov/images/83237/revisiting-bikini-atoll.

24 **had survived the initial explosion:** Amber Dance, "50 years after the blast: Recovery in Bikini Atoll's coral reef," Mongabay.com, May 27, 2008.

24 **commented in wonder, "absolutely pristine":** Rob Taylor, "Coral flourishing at Bikini Atoll atomic test site," Reuters, April 15, 2008.

25 **twenty-eight species of coral were still missing:** Z. T. Richards, M. Beger, S. Pinca, and C. C. Wallace, "Bikini Atoll coral biodiversity resilience five decades after nuclear testing," *Marine Pollution Bulletin* 56, no. 3 (2008): 503–15, doi:10.1016/j.marpolbul.2007:11.018.

25 **and the coral more impressive:** Sam Scott, "What Bikini Atoll looks like today," *Stanford Magazine*, December 2017.

27 **caverns in the south of France:** Jens Lachmund, *Greening Berlin: The Co-Production of Science, Politics, and Urban Nature* (Cambridge, MA, and London: MIT Press, 2013), 167; see also I. Kowarik and A. Langer, "Natur-Park Südgelände: Linking Conservation and Recreation in an Abandoned Railyard in Berlin," *Wild Urban Woodlands* (n.d.), 287–99, doi:10.1007/3-540-26859-6_18.

28 **worth preserving in their own right:** Ingo Kowarik, "Unkraut oder Urwald? Natur der vierten Art auf dem Gleisdreieck," in *Gleisdreieck morgen: Sechs Ideen für einen Park* (Berlin: Bundesgartenschau Berlin GmbH and Berzirksamt Kreuzberg, 1991) 45–55.

28 **any other site in the UK:** The Land Trust, "Canvey Wick," https://thelandtrust.org.uk/space/canvey-wick/?doing_wp_cron=1579053865:2332100868225097656250.

29 **Site of Special Scientific Interest in 2005:** See Patrick Barkham, "Canvey Wick: the Essex 'rainforest' that is home to Britain's rarest insects," *The Guardian*, October 15, 2017.

29 **"from the scene of the explosion":** "Serious Explosion at the Dynamite Works," *Ardrossan and Saltcoats Herald*, May 9, 1884, as reproduced on local-history website Three Towners: https://threetowners.com/ardeer-factory/1884-explosion/.

31 **"that of some ancient woodlands":** G. Barker, "Ecological recombination in urban areas: implications for nature conservation. Proceedings of a workshop held at the Center for Ecology and Hydrology (Monks Wood)," 2000, cited in "Planning for Brownfield Biodiversity: A Best Practice Guide," *Buglife*, Peterborough, 2009.

31 **to turn them into productive farmland:** An excellent cultural history of the concept of wasteland can be found in Vittoria Di Palma, *Wastelands: A History* (New Haven, CT: Yale University Press, 2014).

32 **"as yet uncaptured by language":** Aldo Leopold, "Marshland Elegy," in *A Sand County Almanac, and Sketches Here and There* (Oxford: Oxford University Press, 1949 [1968]), 96.

32 **"clangorous descending spirals":** Leopold, "Marshland Elegy," 96.

34 **left to their own devices:** Harvie compared three methods: unmanaged, traditionally managed (as discussed in my text), and "ecologically managed," as at Addiewell and Oakbank, which she found to be more successful and biodiverse than the traditionally managed sites—but not significantly different from the unmanaged sites like Greendykes; B. Harvie and G. Russell, "Vegetation dynamics on oil-shale bings; implication for management of post industrial sites," *Aspects of Applied Biology* 82 (2007): 57–64.

34 **rambunctious self-seeded community:** "The High Line's planting design is inspired by the self-seeded landscape that grew wild for 25 years after the trains stopped running," www.thehighline.org/gardens, accessed February 9, 2020.

34 **"conditions of the site are artificial":** Erin Eck, "Gardening in the Sky: Wild Inspiration," thehighline.org, November 2, 2017.

34 **"that creeps over the Earth":** Genesis 1:26.

34 **"anthropocentric religion the world has ever seen":** Lynn White Jr., "The Historical roots of our ecologic crisis," *Science* 155, no. 3767, 1203–7, quoted in Carolyn Merchant, *Reinventing Eden* (New York: Routledge, 2004).

36 **their "immaculate and classical nature":** Gallery label: "Derelict Land Art: Five Sisters," 2010, https://www.tate.org.uk/art/artworks/latham-five-sisters-bing-t02072.

36 **reconceptualized as "process sculptures":** Craig Richardson, "Waste to Monument: John Latham's Niddrie Woman: Art & Environment," *Tate Papers* no. 17 (Spring 2012). Available online at: https://www.tate.org.uk/research/publications/tate-papers/17/.

36 **comprised her oversized heart:** Richardson, "Waste to Monument."

38 **clover, ryegrass, fat hen, buttercup:** Richard Mabey, *Weeds: The Story of Outlaw Plants* (London: Profile Books, 2010), 54.

Chapter 2: No Man's Land

43 **in green pencil on a map:** Reportedly a chinagraph pencil. See "Green Line in Beirut Not First in Mideast," *The New York Times*, February 8, 1984, A11.

44 **layer unto the horizon:** Rebecca Solnit, "The Blue of Distance," in *A Field Guide to Getting Lost* (Edinburgh: Canongate, 2005 [2017]), 29.

45 **Lewis observed in his journal:** Lewis journal, May 5, 1805, https://lewisandclark journals.unl.edu/item/lc.jrn.1805-05-05.

45 **"unconcernedly continued to feed":** Lewis journal, May 4, 1805, https://lewisand clarkjournals.unl.edu/item/lc.jrn.1805-05-04.

46 **"horses foaming with sweat":** Lewis journal, August 13, 1805, https://lewisand clarkjournals.unl.edu/item/lc.jrn.1805-08-14.

46 **and nearly two hundred dogs:** A. S. Laliberte and W. J. Ripple, "Wildlife Encounters by Lewis and Clark: A Spatial Analysis of Interactions between Native Americans and Wildlife," *BioScience* 53, no. 10 (2003): 994–1003, doi: 10.1641/0006-3568(2003) 053[0994:WEBLAC]2:0.CO;2.

46 **of around forty-six thousand square miles:** A very interesting statistical analysis of Lewis and Clark's hunting spoils can be found in P. Martin and C. Szuter, "War Zones and Game Sinks in Lewis and Clark's West," *Conservation Biology* 13, no. 1 (1999): 36–45.

47 **thirty-eight thousand square miles:** Martin and Szuter, "War Zones and Game Sinks."

47 **as famine gripped both nations:** A boundary treaty was established in 1825; famine was reported by 1828 and reached a climax in 1831, when war broke out again. See H. Hickerson, "The Virginia deer and intertrival buffer zones in the upper Mississippi Valley," in Anthony Leeds and Andrew P. Vayda, eds., *Man, Culture and Animals* (Washington, D.C.: American Association for the Advancement of Science, 1965), 43–66; also H. Hickerson, *The Chippewa and Their Neighbors: A Study in Ethnohistory* (New York: Holt, Rinehart and Winston, 1970).

47 **"today in many tropical forests":** J. A. McNeely, "Conserving forest biodiversity in times of violent conflict," *Oryx* 37, no. 02 (2003), doi:10.1017/s0030605303000334.

47 **"wild animals are to be found":** Clark journal, August 29, 1806, https://lewisand clarkjournals.unl.edu/item/lc.jrn.1806-08-29#lc.jrn.1806-08-29:01.

54 **the petroleum reek of napalm:** A dramatic eyewitness record of events during the battle for Nicosia airport can be found in F. Henn, "The Nicosia airport incident of 1974: a peacekeeping gamble," *International Peacekeeping* 1, no. 1 (1994): 80–98.

55 **petals, grow in large numbers:** These two plants were specifically surveyed by a separate project: "Cooperation for the Conservation of Endemic Plants in the Buffer Zone," Nature Conservation Unit, Frederick University, Nicosia.

56 **interlude, wild fish stock rebounded:** D. Beare, F. Hölker, G. H. Engelhard, E. McKenzie, and D. G. Reid, "An unintended experiment in fisheries science: a marine area protected by war results in Mexican waves in fish numbers-at-age," *Naturwissenschaften* 97, no. 9 (2010): 797–808, cited in Antonio Uzal, "Rewilding war zones can help heal the wounds of conflict," *The Conversation*, December 18, 2018.

56 **before they were again depleted:** P. Holm, "World War II and the 'Great Acceleration' of North Atlantic Fisheries," *Global Environment* 5, no. 10 (2012): 66–91, cited in Uzal, "Rewilding war zones."

56 **death by the Soviet-era munitions:** Jamie Merril, "Landmine sanctuary: rare leopard finds haven in the lethal legacy of Iran-Iraq war," *The Independent*, December 23, 2014.

56 **great disruption to the ecosystem:** Matthew Teller, "The Falklands penguins that would not explode," BBC News, May 7, 2017.

57   **"red list" of endangered species:** Christian Schwägerl, "Along Scar from Iron Curtain, A Green Belt Rises in Germany," *Yale Environment 360*, April 4, 2011.

59   **themselves set off mines and tripwires:** Testimony from "Kim," featured in *489 Years*, directed by Hayoun Kwon (2016).

Chapter 3: Old Fields

65   **"concentrated on intense everyday work":** R. Taagepera, "Soviet collectivization of Estonian agriculture: the deportation phase," *Soviet Studies* 32, no. 3 (1980): 379–97, doi:10.1080/09668138008411308.

66   **an estimated 245 million square miles:** F. Schierhorn, T. Kastner, T. Kuemmerle, P. Meyfroidt, I. Kurganova, A. V. Prishchepov, et al., "Large greenhouse gas savings due to changes in the post-Soviet food systems," *Environmental Research Letters* 14, no. 6 (2019).

66   **the world's most violent cities:** Jennifer Hanley-Giersch, "The Baltic States and the North Eastern European criminal hub," *ACAMS Today* 8, no. 4, (2009). *ACAMS Today* is a publication of the Association of Certified Anti-Money Laundering Specialists.

69   **as evolution is now to general biology:** Marcel Rejmánek and Kristina P. Van Katwyk, "Old-field succession: A bibliographic review, 1901–1991," Section of Evolution and Ecology, University of California, Davis, http://botanika.prf.jcu.cz/suspa/pdf/BiblioOF.pdf.

70   **of the country in 1920:** Adele Johanson, "The state of Estonia's forest is the best in a century," *Postimees*, June 7 2017.

70   **fall of the Soviet Union:** Calculated by combining the figures for "forest" and "other wooded land," found on page 9 of The Food and Agriculture Organization of the United Nations (FAO), "Global Forest Resources Assessment 2015: Country Profile—Estonia."

70   **forest has "naturally regenerated":** Of that forest, 3 percent can be considered old growth; the FAO quoted in Rachel Fritts, "Estonia's trees: Valued resource or squandered second chance?" Mongabay.com, October 20, 2017.

70   **eastern Europe and European Russia alone:** Adam Voiland, "Changing Forest Cover Since the Soviet Era," NASA Earth Observatory, July 16, 2015, https://earthobservatory.nasa.gov/images/86221/changing-forest-cover-since-the-soviet-era.

70   **man-made carbon sink in history:** Michael Slezak, "Fall of USSR locked up world's largest store of carbon," *New Scientist*, October 2, 2013.

70   **attempted to quantify the impact:** Schierhorn, Kastner, et al., "Large greenhouse gas savings."

71   **the abandonment of farmland alone:** Slezak, "Fall of USSR locked up world's largest store of carbon."

73 **"part" of this missing carbon:** "In sum, our analyzes suggest that cropland abandonment in the [former Soviet Union] may explain a considerable part of the global
residual terrestrial C sink since 1991." In Schierhorn, Kastner, et al., "Large greenhouse gas savings."

73 **than in the tropics or subarctic:** Gabriel Popkin, "The hunt for the world's missing
carbon," *Nature*, June 30, 2015.

73 **"Trees have already been invented":** S. B. Hecht, K. Pezzoli, and S. Saatchi, "Trees
have already been invented: carbon in woodlands," *Collabra* 2, no. 1 (2016), doi:10
.1525/collabra.69.

73 **a "dramatic and ongoing" increase:** Viki A. Cramer and Richard J. Hobbs, eds., *Old
Fields: Dynamics and Restoration of Abandoned Farmland* (Washington, D.C.: Island
Press, 2007), 2.

73 **a global analysis in 1999:** N. Ramankutty and J. A. Foley, "Estimating Historical
Changes in Global Land Cover: Croplands from 1700 to 1992," *Global Biogeochemical
Cycles* 13 (1999): 997–1027, doi:10.1029/1999GB900046.

74 **woodlands were standing by 1850:** G. G. Whitney and W. C. Davis, "From Primi-
tive Woods to Cultivated Woodlots: Thoreau and the Forest History of Concord,
Massachusetts," *Forest & Conservation History* 30, no. 2 (1986): 70–81, doi:10.2307
/4004930.

75 **woodpeckers have rebounded too:** Colin Nickerson, "New England sees a return of
forests, wildlife," *The Boston Globe*, August 31, 2013.

75 **book than it did in 1845:** This point made in Robert Sattelmeyer, "Depopulation,
Deforestation, and the Actual Walden Pond," in R. J. Schneider, ed., *Thoreau's Sense
of Place: Essays in American Environmental Writing* (Iowa City: University of Iowa Press,
2000), 235–43.

75 **every year between 1910 and 1979:** Michael Williams, "Dark Ages and Dark Areas:
Global Deforestation in the Deep Past," *Journal of Historical Geography* 26 (2000):
28–46, doi:10.1006/jhge.1999:0189.

75 **abandoned between 2000 and 2030:** Graham Lawton, "The Call of Rewilding,"
New Scientist, October 13, 2018.

75 **tripled its forest area:** José M. Rey Benayas, "Rewilding: As farmland and villages
are abandoned, forests, wolves and bears are returning to Europe," *The Conversation*,
July 2, 2019.

75 **brown bears have spiked:** M. Enserink and G. Vogel, "Wildlife Conservation: the
Carnivore Comeback," *Science* 314, no. 5800 (2006): 746–49, doi:10.1126/science.314
:5800:746.

76 **860,000 square miles, since 1982:** X.-P. Song, M. C. Hansen, S. V. Stehman, P. V.
Potapov, A. Tyukavina, E. F. Vermote, and J. R. Townshend, "Global land change
from 1982 to 2016," *Nature* 560 (2018): 639–43, doi:10.1038/s41586-018-0411-9.

76 **and growing in the final third:** "Global Forest Resources Assessment," Food and
Agriculture Organization of the United Nations (FAO), Rome, 2015. NB: this paper,

Notes

published three years before Song et al.'s "Global land change from 1982 to 2016," states that global forest cover declined 3 percent between 1990 and 2015. The contradiction is likely due to varying methodologies. Song et al.'s findings, being based on thirty-five years of satellite data, published in a respected journal, and acclaimed as "the most comprehensive picture ever made of the changing use of land" (*The Independent*, August 10, 2018), seem to me well founded.

76 **approaching forest transition:** R. J. Keenan, G. A. Reams, F. Achard, J. V. de Freitas, A. Grainger, and E. Lindquist, "Dynamics of global forest area: Results from the FAO Global Forest Resources Assessment 2015," *Forest Ecology and Management*, 352 (2015), 9–20, doi:10.1016/j.foreco.2015.06.014.

77 **lifetimes a reversal of deforestation:** P. Meyfroidt and E. F. Lambin, "Global Forest Transition: Prospects for an End to Deforestation," *Annual Review of Environment and Resources* 36, no. 1 (2011): 343–71, doi:10.1146/annurev-environ-090710-143732.

77 **rainforest are deserted each year:** R. A. Houghton, D. L. Skole, C. A. Nobre, J. L. Hackler, K. T. Lawrence, and W. H. Chomentowski, "Annual fluxes of carbon from deforestation and regrowth in the Brazilian Amazon," *Nature* 403, no. 6767 (2000): 301–4, doi:10.1038/35002062; also Hecht et al., "Trees have already been invented."

77 **two-thirds of the world's forest:** In the region of 68 percent, calculated thusly: most of the world's forest is natural forest (93 percent) and most natural forest falls into the category of "other naturally regenerated forest" (74 percent), http://fao.org/3/a-i4793e.pdf.

78 **it took three days to traverse:** Minhaj ud-Din Juzjani, *Tabakat-i-Nasiri*, trans. Major H. G. Raverty, 3rd reprint (Kolkata: Asiatic Society, 2010), 965, quoted in Timothy May, ed., *The Mongol Empire: A Historical Encyclopedia*, vol. 1 (Santa Barbara, CA: ABC-CLIO, 2017), 219.

78 **as millions fled to the south:** K. G. Deng, *China: Tang, Song and Yuan Dynasties, The Oxford Encyclopedia of Economic History*, vol. 2, ed. J. Mokyr (Oxford: Oxford University Press, 2003), 423–28, cited in William F. Ruddiman and Ann G. Carmichael, "Pre-Industrial Depopulation, Atmospheric Carbon Dioxide, and Global Climate," Interactions Between Global Change and Human Health Pontifical Academy of Sciences, *Scripta Varia* 106, Vatican City (2006).

78 **In a 2011 paper:** J. Pongratz, K. Caldeira, C. H. Reick, and M. Claussen, "Coupled climate–carbon simulations indicate minor global effects of wars and epidemics on atmospheric CO_2 between ad 800 and 1850," *The Holocene* 21, no. 5 (2011): 843–51, doi:10.1177/0959683610386981.

79 **"were raining down from heaven":** Quoted in M. Wheelis, "Biological Warfare at the 1346 Siege of Caffa," *Emerging Infectious Diseases* 8, no. 9 (2002): 971–75, https://dx.doi.org/10.3201/eid0809:010536.

79 **"or flee or escape from them":** Whelis, "Biological Warfare at the 1346 Siege of Caffa."

80   **north at around 2.5 miles a day:** C. J. Duncan and S. Scott, "What caused the Black Death?" *Postgraduate Medical Journal* 81, no. 955 (2005).

80   **all settlements were abandoned:** Williams, "Dark Ages and Dark Areas."

80   **he wrote, fields were lying uncultivated:** Quoted in Francis Aidan Gasquet, *The Great Pestilence* (London: Simpkin Marshall, Hamilton, Kent & Co., 1893; Gutenberg Project, 2014), 51.

80   **all "overgrown with brambles and bushes":** Michael Williams, *Deforesting the Earth: from Prehistory to Global Crisis* (Chicago: University of Chicago Press, 2003), 136.

80   **forests today date from this period:** Williams, *Deforesting the Earth*, 136.

81   **American paleoclimatologist William Ruddiman:** W. F. Ruddiman, "The Anthropogenic Greenhouse Era Began Thousands of Years Ago," *Climatic Change* 61, no. 3 (2003): 261–93, doi:10.1023/b:clim.0000004577:17928.fa.

81   **on the newly abandoned farmland:** Ruddiman, "The Anthropogenic Greenhouse Era."

81   **gigatons of carbon in total:** The Food and Agriculture Organization of the United Nations, *The State of the World's Forests 2018—Forest pathways to sustainable development* (Rome, 2018).

82   **touching down in 1492 and 1650:** W. M. Denevan, "The Pristine Myth: The Landscape of the Americas in 1492," *Annals of the Association of American Geographers* 82, no. 3 (1992): 369–85, doi:10.1111/j.1467-8306:1992.tb01965.x.

82   **individuals were still alive:** Denevan, "The Pristine Myth," 369–85.

82   **"greatest demographic disaster ever":** Denevan, "The Pristine Myth," 369–85.

82   **eminent geographer William Denevan:** H. C. Heaton, ed., and Bertram T. Lee, trans., *The Discovery of the Amazon: According to the Account of Friar Gaspar de Carvajal and Other Documents* (New York: American Geographical Society, 1934), 198. An editor's note on page 47 of the same book suggests that during the fifteenth century, a league = 3.66 miles.

83   **Manhattan is 13.4 miles long:** The same comparison is made in David Wilkinson, "Amazonian Civilization?" *Comparative Civilizations Review* 74, no. 74, article 7 (2016).

83   **"so bright that they astonish":** Heaton and Lee, *Discovery of the Amazon*, 201.

84   **elaborate drainage and irrigation systems:** W. E. Doolittle, "Agriculture in North America on the Eve of Contact: A Reassessment," *Annals of the Association of American Geographers* 82, no. 3 (1992): 386–401, doi:10.1111/j.1467-8306.1992.tb01966.x.

84   **in the space of a century:** Simon L. Lewis and Mark A. Maslin, "Defining the Anthropocene," *Nature* 519 (2015): 171–80; see also A. Koch, C. Brierley, M. M. Maslin, and S. L. Lewis, "Earth system impacts of the European arrival and Great Dying in the Americas after 1492," *Quaternary Science Reviews* 207 (2019): 13–36, doi:10.1016/j.quascirev.2018.12.004.

Notes

Chapter 4: Nuclear Winter

88 forgotten it ever happened: Iurii Shcherbak, *Chernobyl: A Documentary Story* (New York: St. Martin's Press, 1989).

88 himself, like a rainbow: Youri Korneev, quoted in the documentary *The Battle of Chernobyl*, directed by Thomas Johnson (2006).

92 one every ten thousand years: *Soviet Life* magazine, February 1986, cited in "Odds of a meltdown 'one in 10,000 years' Soviet official says," Associated Press, April 29, 1986.

92 45 percent chance of meltdown: "The Next Nuclear Meltdown," *The New York Times*, May 8, 1985, A26.

93 an estimated 8,800 square miles: Commissioner Greta Joy Dicus, "Presentation to the joint meeting of American Nuclear Society," U.S. Nuclear Regulatory Commission, Washington, D.C. Section and Health Physics Society, Baltimore-Washington Chapter, January 16, 1997.

93 fenced off as a "radiation reserve": Fred Pearce, "Exclusive: First visit to Russia's secret nuclear disaster site," *New Scientist*, December 7, 2016.

93 during the Second World War: C. M. Heeb, "Iodine-131 releases from the Hanford Site, 1944 through 1947" (Oak Ridge, TN: US Office of Scientific and Technical Information, 1993).

97 wolves had increased sevenfold: T. G. Deryabina, S. V. Kuchmel, L. L. Nagorskaya, T. G. Hinton, J. C. Beasley, A. Lerebours, and J. T. Smith, "Long-term census data reveal abundant wildlife populations at Chernobyl," *Current Biology* 25, no. 19 (2015): R811–R826, doi:10.1016/j.cub.2015.08.017.

99 "destruction by greedy developers": James Lovelock, "We need nuclear power, says the man who inspired the Greens," *The Daily Telegraph*, August 16, 2001.

100 fungi, wood ash, human teeth: For an overview of the impacts upon Chernobyl wildlife suitable for the general reader, I recommend Mary Mycio, *Wormwood Forest: A Natural History of Chernobyl* (Washington, D.C.: Joseph Henry Press, 2005).

100 the most contaminated areas: A very good summary of their findings can be found in T. A. Mousseau and A. P. Møller, "Genetic and Ecological Studies of Animals in Chernobyl and Fukushima," *Journal of Heredity* 105, no. 5 (2014): 704–9, doi:10.1093/jhered/esu040.

100 at a worryingly slow rate: T. A. Mousseau, G. Milinevsky, J. Kenney-Hunt, and A. P. Møller, "Highly reduced mass loss rates and increased litter layer in radioactively contaminated areas," *Oecologia* 175, no. 1 (2014): 429–37, doi:10.1007/s00442-014-2908-8.

100 densely packed from poor growth: T. A. Mousseau, S. M. Welch, I. Chizhevsky, O. Bondarenko, G. Milinevsky, et al., "Tree rings reveal extent of exposure to ionizing radiation in Scots pine *Pinus sylvestris*," *Trees* 27, no. 5 (2013): 1443–53, doi:10.1007/s00468-013-0891-z.

101 in extremely contaminated areas: S. C. Webster, M. E. Byrne, S. L. Lance, C. N. Love, T. G. Hinton, D. Shamovich, and J. C. Beasley, "Where the wild things are: influence of radiation on the distribution of four mammalian species within the Chernobyl Exclusion Zone," *Frontiers in Ecology and the Environment* 14, no. 4 (2016): 185–90, doi:10.1002/fee.1227.

102 with anything approaching certainty: "To date, there has been no persuasive evidence of any other health effect in the general population that can be attributed to radiation exposure," United Nations Scientific Committee on the Effects of Atomic Radiation, 2008, quoted in report, "Sources and Effects of Ionizing Radiation," UN, New York, 2011. Cataracts and protracted exposure to low level radiation discussed in World Health Organization release, "1986–2016: Chernobyl at 30," April 25, 2016.

102 that has settled the question: See Kate Brown, *Manual for Survival: A Chernobyl Guide to the Future* (New York: W. W. Norton & Co., 2019) for a plangent warning of the costs to life and public health that may have gone and may still be going unreported and unstudied.

PART TWO: THOSE WHO REMAIN
Chapter 5: The Blight

114 thought to lie vacant in Detroit: For U.S. Postal Service figures for September 2019, see "Detroit Vacancies Decline Over Long-Term, Slow Uptick Recently in Numbers," *Drawing Detroit*, November 11, 2019, http://www.drawingdetroit.com /detroit-vacancies-decline-over-long-term-slow-uptick-recently-in-numbers/.

114 nineteen thousand in five years: Corey Williams, Mike Schneider, and Angeliki Kastanis, "Analysis: Detroit will be toughest US city to count population for 2020 Census," Associated Press, December 12, 2019.

115 one hundred and thirty-nine square miles lie vacant: Detroit Future City, *139 Miles: Detroit Future City* (Detroit: Inland Press, 2017), 71.

119 "Must have like death that we have": John Webster, *The Duchess of Malfi* (London: Methuen & Co., 1623), 5.3.16–18. References are to act, scene, and line.

120 resources, for $0.45 a pound: Scott Hocking, a Detroit-based installation artist who builds beautiful, complex structures from materials amassed from abandoned industrial sites—an egg of marble fragments inside Michigan Central Station, a ziggurat of creosoted blocks on a factory floor—has written and spoken very interestingly about this nocturnal, unofficial industry; see https://www.detroitresearch .org/pictures-of-a-city-scrappers/ and https://www.tbdmag.com/detroit-artist-scott -hocking/.

121 as "outwards signs of blight": Violet Ikonomova, "Despite demolition efforts, blight spreads undetected throughout Detroit's neighborhoods," *Detroit Metro Times*, November 14, 2018.

Notes

123 **a huge drop in house prices:** Allan Mallach, "The Empty House Next Door: Understanding and Reducing Vacancy and Hypervacancy in the United States," Lincoln Institute for Land Policy, May 2018, 19.

123 **buildings, specifically violent assaults:** "Abandoned buildings: magnets for crime?" *Journal of Criminal Justice* 21, no. 5 (1993): 481–95, doi:10.1016/0047-2352(93)90033-j. See also C. C. Branas, D. Rubin, and W. Guo, "Vacant Properties and Violence in Neighborhoods," *ISRN Public Health*, 2012, 1–23, doi:10.5402/2012/246142.

123 **city, according to the FBI:** "Crime drops in Detroit, FBI's most violent big city; search your community's stats," *Detroit News*, September 30, 2019.

123 **roofs of abandoned skyscrapers:** Falcons have been reported nesting on the roof of the city's Lee Plaza, among others.

125 **the lungs of sleeping Glaswegians:** "No city for old men," *The Economist*, August 25, 2012.

125 **"tumor, blight grows back":** Glenda D. Price, Linda Smith, and Dan Gilbert, "A Message from the Chairs," in *Detroit Blight Removal Task Force Plan*, May 27, 2014.

126 **"men are afraid of the light":** *Detroit Blight Removal Task Force Plan.*

126 **George, a former insurance salesman:** Clare Pfeiffer Ramsay, "Who Ya Gonna Call?" *Model D Media*, October 25, 2005.

127 **"in Detroit, Michigan, 1949":** Constance M. King, interviewed by Ellen Piligian on behalf of the author, December 2019.

131 **"But it is home. This is home":** John Carlisle, "Is this the end of Delray?" *Detroit Free Press*, December 11, 2017.

132 **paper, soon to be published:** David Walsh, Gerry McCartney, Chik Collins, Martin Taulbut, and G. David Batty, "History, politics and vulnerability: explaining excess mortality in Scotland and Glasgow," Glasgow Center for Population Health, May 2016.

133 **"make one long for disorder":** *L'architecte*, Paris, September 1925, quoted in Le Corbusier, *The City of Tomorrow and Its Planning* trans. Frederick Etchells (Dover Publications, 1987; original translation: 1929), 133.

Chapter 6: Days of Anarchy

134 **"assumes the polish of a mirror":** Quoted in Arthur S. Lefkowitz, *George Washington's Indispensable Men: Alexander Hamilton, Tench Tilghman, and the Aides-de-Camp Who Helped Win American Independence* (Lanham, MD: Rowman & Littlefield, 2018), 178.

134 **drifted away on the breeze:** *The Life and Correspondence of James McHenry* (Cleveland: The Burrows Brothers Company, 1907), 22.

135 **from the base of the oak:** Lefkowitz, *George Washington's Indispensable Men.*

136 **"Bethlehem of capitalism":** quoted in Peter Applebome, "Paterson, the Yellowstone of the East?" *The New York Times*, February 12, 2006.

136 **"system . . . the forgotten zone":** Wheeler Antabanez, *Weird N.J. Presents: Nightshade on the Passaic* (Bloomfield: Weird N.J., 2008).

137 "and the present pouring down": William Carlos Williams, *Paterson* (New York: New Directions, 1946–1958), bk. 3, pt. 3.

142 "would be a true *Inferno*": Paul Mariani, *William Carlos Williams: A New World Naked* (1981; San Antonio: Trinity University Press, 1990), 419.

142 "it's a flower to the mind too": Quoted in William Carlos Williams's *Paterson*, bk. 5, pt. 1.

143 he says, as introduction: "Cesar" is a pseudonym.

149 by the color of the Passaic: Chris Sturm and Nicholas Dickerson, "The Power of the Passaic: Paterson's Birth and Rebirth Along the River," in *Ripple Effects: The State of Water Infrastructure in New Jersey Cities and Why it Matters* (Trenton: New Jersey Future, 2014), 41.

149 "out hot, / swirling, bubbling": William Carlos Williams, *Paterson*, bk. 1, pt. 3.

150 "at once from the poisonous water": Anon., "Turned white paper black: polluted Passaic water that Jersey City drinks," *The New York Times*, August 26, 1894, 17.

150 the sewerage commission reported: Passaic Valley Sewerage Commission, *Report of the Passaic Valley Sewerage Commission upon the general system of sewage disposal for the valley of the Passaic River, and the prevention of pollution thereof* (Newark, NJ: John E. Rowe & Son, 1897).

150 the Passaic River's total flow: N. F. Brydon, *The Passaic River: Past, Present, Future* (New Brunswick, NJ: Rutgers University Press, 1974); cited in Victor Onwueme and Huan Feng, "Risk characterization of contaminants in Passaic River sediments, New Jersey," *Middle States Geographer* 39 (2006).

150 "hair and eyebrows of watchmen": Anon., "The Passaic river fire," *The New York Times*, June 6, 1918.

151 "swillhole in all of Christendom": William Carlos Williams, *In the American Grain* (1925; New York: New Directions, 2009).

PART THREE: THE LONG SHADOW

Chapter 7: Unnatural Selection

156 he would later remark: "Oral history interview with Robert Smithson," Smithsonian Archives of American Art, July 14–19, 1972.

157 "ruins before they are built": Robert Smithson, "The Monuments of Passaic," *Artforum*, December 1967, 52–57.

157 "abandoned set of futures": Smithson, "The Monuments of Passaic."

159 a time-lapse of its ruination: Aerial photographs charting the island's degeneration can be found on the Library of Congress website, https://www.loc.gov/item/ny1414/.

Notes

160 **the sludge beneath them:** Jonah Owen Lamb, "The Ghost Ships of Coney Island Creek," *The New York Times,* August 6, 2006. New York State Department of Environmental Conservation confirmed in private correspondence with the author that there are no plans to lift the wrecks, and declined to speculate further on the nature of the contaminants.

161 **"New Jersey's biggest crime scene":** Ted Sherman, "Massive, $1.7 billion environmental cleanup of Passaic River proposed by EPA," NJ Advance Media for NJ.com, April 11, 2014.

162 **thirty thousand gallons of it a day:** Mary Bruno, *An American River: From Paradise to Superfund, Afloat on the Passaic River* (Vashon, WA: DeWitt Press, 2012), 68.

163 **they are virtually indestructible:** Some dechlorination of PCBs over time, in highly contaminated environments, has been identified. The tiny anaerobic bacteria *Dehalococcoides mccartyi* was identified in 2007 as the likely culprit, raising the hope that scientists may develop enzymatic or bioremediative methods of environment restoration. For TCDD, incineration is currently the best method of disposal during, for example, Superfund site clean-ups. When PCBs are incinerated, they can produce dioxins.

163 **be at least a century:** S. Sinkkonen and J. Paasivirta, "Degradation half-life times of PCDDs, PCDFs and PCBs for environmental fate modeling," *Chemosphere* 40, nos. 9–11 (2000): 943–49, doi:10.1016/s0045-6535(99)00337-9.

163 **it as "virtually nonbiodegradable":** E.g., United States Department for Agriculture's Agricultural Research Service, "Monitoring Dioxins," *Agricultural Research* 49, no. 1 (2001): 14–15.

164 **"oily refuse into the waters":** Eugene G. Blackford, "Report on an Oyster Investigation in New York with the Steamer Lookout," quoted in Bonnie J. McCay, *Oyster Wars and the Public Trust: Property, Law, and Ecology in New Jersey History* (Tucson: University of Arizona Press, 1998), 156.

167 **the past twenty-five years:** Damian Carrington, "UK killer whale died with extreme levels of toxic pollutants," *The Guardian,* May 2, 2017.

168 **classified as hazardous waste:** Marla Cone, "Pollutants drift north, making Inuits' traditional diet toxic," *Los Angeles Times,* January 13, 2004.

168 **had cancerous tumors:** W. K. Vogelbein, J. W. Fournie, P. A. Vanveld, and R. J. Huggett, "Hepatic neoplasms in the mummichog *Fundulus heteroclitus* from a creosote-contaminated site" (1990), cited in R. T. Di Giulio and B. W. Clark, "The Elizabeth River Story: A Case Study in Evolutionary Toxicology," *Journal of Toxicology and Environmental Health, Part B, Critical Reviews* 18, no. 6 (2015): 259–98, doi:10.1080/15320383:2015:1074841.

169 **the killifish had done it:** N. M. Reid, D. A. Proestou, B. W. Clark, W. C. Warren, J. K. Colbourne, et al., "The genomic landscape of rapid repeated evolutionary adaptation to toxic pollution in wild fish," *Science* 354, no. 6317 (2016): 1305–8, doi:10.1126/science.aah4993.

Notes

169 **harmful effects of PCBs:** I. Wirgin, N. K. Roy, M. Loftus, R. C. Chambers, D. G. Franks, and M. E. Hahn, "Mechanistic Basis of Resistance to PCBs in Atlantic Tomcod from the Hudson River," *Science* 331, no. 6022 (2011): 1322–25, doi:10.1126/science.1197296.

170 **mile of the city a year:** Judith Hooper, *Of Moths and Men: Intrigue, Tragedy and the Peppered Moth* (2002; London: HarperCollins, 2012), ebook loc. 299.

171 **of the peppered moth:** Hooper, *Of Moths and Men,* loc. 285.

171 **lichen-free, it flourished:** Bernard Kettlewell's studies of the 1950s, which demonstrated selective predation at work, were recreated later by Michael Marjerus after questions were raised over Kettlewell's original methodology (see Hooper, *Of Moths and Men,* for an overview of the controversy). Majerus's results were published posthumously: L. M. Cook, B. S. Grant, I. J. Saccheri, and J. Mallet, "Selective bird predation on the peppered moth: the last experiment of Michael Majerus," *Biology Letters* 8, no. 4 (2012): 609–12, doi:10.1098/rsbl.2011.1136.

172 **and maybe even millennia:** Stephen R. Palumbi, "Humans as the World's Greatest Evolutionary Force," *Science* 293, no. 5536 (2001): 1786–90.

172 **rate of 2 to 6 percent:** Anna M. Whitehouse, "Tusklessness in the elephant population of the Addo Elephant National Park, South Africa," *Journal of Zoology* 257, no. 2 (2002): 249–54, doi:10.1017/S0952836902000845.

172 **species by 300 percent:** Chris T. Darimont, Stephanie M. Carlson, Michael T. Kinnison, Paul C. Paquet, Thomas E. Reimchen, and Christopher C. Wilmers, "Human predators outpace other agents of trait change in the wild," *Proceedings of the National Academy of Sciences* 106, no. 3 (2009): 952–54, doi:10.1073/pnas.0809235106.

173 **blood to that of birds:** K. Byrne and R. Nichols, "*Culex pipiens* in London Underground tunnels: differentiation between surface and subterranean populations," *Heredity* 82 (1999): 7–15, doi:10.1038/sj.hdy.6884120.

176 **less than a third:** C. A. Clarke, F. M. M. Clarke, and H. C. Dawkins, "*Biston betularia* (the peppered moth) in West Kirby, Wirral, 1959–89: updating the decline in f. *carbonaria,*" *Biological Journal of the Linnean Society* 39, no. 4 (1990): 323–26, doi:10.1111/j.1095-8312:1990.tb00519.x.

Chapter 8: Forbidden Forest

179 **the landscape like a hose:** "like a garden hose," Corporal Stephane, quoted in Alistair Horne, *The Price of Glory: Verdun 1916* (1926; London: Penguin, 1993), ebook loc. 1416.

179 **There's nothing left alive:** Horne, *Price of Glory,* loc. 1465.

180 **years of natural erosion:** Jean-Paul Amat, quoted in AFP article, "La forêt de Verdun, écrin vert créé par la guerre," *La Croix,* May 2016.

181 **"earth itself is corpselike":** Henri Barbusse, *Under Fire,* trans. Robin Buss (New York: Penguin Books, 2004), 5, 7, 138, 248. Quoted in Tait Keller, "Destruction of

the Ecosystem," *International Encyclopedia of the First World War,* October 8, 2014, https://encyclopedia.1914-1918-online.net/article/destruction_of_the_ecosystem.

183 **holders of power lines:** Jean-Paul Amat, "Guerre et milieux naturels: les forêts meur-tries de l'Est de la France, 70 ans après Verdun," *L'Espace Géographique* no. 3 (1987): 217–33.

185 **two hundred thousand of them:** J. Forget, "La reconstitution forestière de la zone rouge dans las Meuse," in *Bulletin de la Société Des Lettres, Sciences et Arts de Bar-le-Duc et du Musée de Géographie* (1928), 121–31, cited in Daniel Hubé, "3. La Place à Gaz de la forêt de Spincourt, une zone industrielle toxique," *Mission Centenaire,* April 19, 2018, http://centenaire.org/fr/espace-scientifique/3-la-place-gaz-de-la-foret-de-spincourt -une-zone-industrielle-toxique.

186 **Then, finally, in 1928:** Hugues Thouin, Lydie Le Forestier, Pascale Gautret, Dan-iel Hube, Valérie Laperche, et al., "Characterization and mobility of arsenic and heavy metals in soils polluted by the destruction of arsenic-containing shells from the Great War," *Science of the Total Environment* 550 (2016): 658–69, 10.1016/j.scitotenv .2016.01.111.

187 **In 2007, the German scientists:** T. Bausinger, E. Bonnaire, and J. Preuss, "Expo-sure assessment of a burning ground for chemical ammunition on the Great War battlefields of Verdun," *Science of the Total Environment* 382, nos. 2–3 (2007): 259–71. doi: 10.1016/j.scitotenv.2007.04.029.

188 **metals were added separately:** N. G. Nesvetaylova, "Geobotanical Investigations in Prospecting for Ore Deposits," *International Geology Review* 3, no. 7 (1961): 609–18, doi:10.1080/00206816109473622.

188 **contents of the underworld:** D. P. Malyuga, *Biogeochemical Methods of Prospecting* (Moscow: Izd. Akad. Nauk SSSR, 1959; New York: Consultant's Bureau, 1964), quoted in H. L. Cannon, "The Use of Plant Indicators in Ground Water Surveys, Geologic Mapping, and Mineral Prospecting," *Taxon* 20, nos. 2–3 (1971): 227–56, doi:10.2307/1218878.

189 **as a means of prospecting:** Robert Temple, *The Genius of China* (1986; London: Prion Books, 1999), 159–60.

190 **2,000 parts per million:** Cannon, "Use of Plant Indicators in Ground Water Sur-veys."

190 **Africa; in Russia, boron:** Cannon, "Use of Plant Indicators in Ground Water Sur-veys."

195 **has been an exponential growth:** L. A. B. Novo, P. M. L. Castro, P. Alvarenga, and E. F. da Silva, "Phytomining of Rare and Valuable Metals," in *Phytoremediation: Man-agement of Environmental Contaminants,* vol. 5 (Springer, 2017): 469–86, doi:10.1007 /978-3-319-52381-1_18.

195 **sites have been reported:** Z. He, J. Shentu, X. Yang, V. C. Baligar, T. Zhang, and P. J. Stoffella, "Heavy Metal Contamination of Soils: Sources, Indicators, and Assess-ment," *Journal of Environmental Indicators* 9 (2015): 17–18.

Notes

196 **follow-up paper in 2016:** Thouin et al., "Characterization and mobility of arsenic and heavy metals in soils."

197 **"from divers examples":** John Webster, *Metallographia, or, an history of metals* (Kettilby, 1671), 47–48.

197 **"fruits, and thin plates":** Webster, *Metallographia.*

198 **"Our death, the Tree of Knowledge":** John Milton, *Paradise Lost* (1667; Oxford: Oxford University Press, 1813), bk. 5, lines 218–21.

199 **dress and onto the floor:** Alison Matthews David, *Fashion Victims: The Dangers of Dress Past and Present* (London: Bloomsbury, 2015), excerpted online at https://pictorial .jezebel.com/the-arsenic-dress-how-poisonous-green-pigments-terrori-1738374597.

Chapter 9: Alien Invasion

203 **containing two thousand further species:** "The botanical gardens are spread over some 300 ha, and originally consisted of 20 plantation blocks, divided into 141 compartments; these compartments vary in shape and size, from 0.1 to 7 ha, and originally contained almost 2,000 species plots of varying size . . . the collections were subject to near abandon between 1948 and 1993. Accordingly . . . approximately only one-third of species and plots remain." From P. J. Greenway, "Report of a botanical survey of the indigenous and exotic plants in cultivation at the East African Agricultural Research Station, Amani, Tanganyika Territory," unpublished typescript, Royal Botanic Gardens, Kew, 1934; cited in P. E. Hulme, D. F. R. P. Burslem, W. Dawson, E. Edward, J. Richard, and R. Trevelyan, "Aliens in the Arc: Are Invasive Trees a Threat to the Montane Forests of East Africa?" in L. Foxcroft, P. Pyšek, D. Richardson, and P. Genovesi, eds., *Plant Invasions in Protected Areas. Invading Nature—Springer Series in Invasion Ecology, vol.* 7 (Dordrecht: Springer, 2013): 145–65, doi:10.1007/978-94-007-7750-7_8.

203 **over one hundred million sheep:** Fred Pearce, *The New Wild* (London: Icon Books, 2016), 33.

203 **one sailor panted:** U.S. Navy captain David Porter, quoted in Paul Chambers, *A Sheltered Life: the Unexpected History of the Giant Tortoise* (Oxford: Oxford University Press, 2006), 100.

204 **foundered in fits of fever:** Will McGuire, "Long live the weeds," Kew Gardens, London, November 30, 2018: https://www.kew.org/read-and-watch/long-live-the -weeds.

204 **"The finch, the sparrow, and the lark":** William Shakespeare, *A Midsummer Night's Dream,* 3.1.61.

204 **"Nothing but 'Mortimer'":** William Shakespeare, *Henry IV,* 1.3.224.

205 **"the continent from the sky":** Sir Mick Gillies, quoted in P. Wenzel Geissler and Ann H. Kelly, "Field Station as Stage: Re-Enacting Scientific Work and Life in Amani, Tanzania," *Social Studies of Science* 46, no. 6 (December 2016): 912–37, doi .org/10.1177/0306312716650045.

Notes

208 **recorded again until 1962:** D. W. Holt, R. Berkley, C. Deppe, P. Enríquez Rocha, J. L. Petersen, J. L. Rangel Salazar, K. P. Segars, and K. L. Wood, "Usambara Eagle-owl (*Bubo vosseleri*)," in J. del Hoyo, A. Elliott, J. Sargatal, D. A. Christie, and E. de Juana, eds., *Handbook of the Birds of the World Alive* (Barcelona: Lynx Edicions).

208 **and never seen again:** T. Barbour and A. Loveridge, "A comparative study of the herpetological faunae of the Uluguru and Usambara mountains, Tanganyika territory with descriptions of new species," printed for the Museum of Comparative Zoology, Harvard University, Cambridge, MA, 1928. Page 261 describes discovery.

212 **lurking in his trouser cuffs:** Richard Mabey, *Weeds: The Story of Outlaw Plants* (London: Profile Books, 2010), 30.

213 **rampant through old-growth forest:** W. Dawson, A. S. Mndolwa, D. F. R. P. Burslem, and P. E. Hulme, "Assessing the risks of plant invasions arising from collections in tropical botanical gardens," *Biodiversity and Conservation* 17, no. 8 (2008): 1979–95, https://doi.org/10:1007/s10531-008-9345-0.

213 **trees in the country:** Woodland Trust, "Key tree pests and diseases," https://www .woodlandtrust.org.uk/trees-woods-and-wildlife/tree-pests-and-diseases/key -tree-pests-and-diseases/.

213 **on an imported tree:** Mark Kinver, "Ash dieback: Killer tree disease set to cost UK £15bn," BBC News, May 6, 2019.

213 **able to withstand its attacks:** Forest Research, "Ash dieback (Hymenoscyphus fraxineus)," https://www.forestresearch.gov.uk/tools-and-resources/pest-and-disease -resources/ash-dieback-hymenoscyphus-fraxineus/.

213 **piqued by the "astounding":** Charles Darwin, private correspondence with J. D. Hooker, dated January 25, 1862, https://www.darwinproject.ac.uk/letter/DCP -LETT-3411.xml.

213 **"formed of snow-white wax":** From Charles Darwin's *On the various contrivances by which British and foreign orchids are fertilized by insects...*, quoted in Martin Brinkworth and Friedel Weinert, eds., *Evolution 2.0: Implications of Darwinism in Philosophy and the Social and Natural Sciences* (Heidelberg: Springer-Verlag, 2012), 96.

213 **"what insect can suck it":** Darwin, letter to J. D. Hooker, January 25, 1862.

214 **slipped right into place:** The full fascinating story is told in J. Arditti, J. Elliott, I. J. Kitching, and L. T. Wasserthal, "'Good Heavens what insect can suck it'— Charles Darwin, *Angraecum sesquipedale* and *Xanthopan morganii praedicta*," *Botanical Journal of the Linnean Society* 169, no. 3 (2012): 403–32, doi:10.1111/j.1095-8339:2012 :01250.x.

214 **to one another's tines:** Pearce, *The New Wild*, 179–92, offers a very effective potted history of the ideas of ecosystems, niches, and coevolution, as well as current rival schools of thought. Pearce's writing has strongly influenced my thinking in this area.

214 **"believe what he can"**: This statement is worth viewing in full context: "With respect to the theological view of the question. This is always painful to me. I am bewildered. I had no intention to write atheistically. But I own that I cannot see as plainly as others do, and as I wish to do, evidence of design and beneficence on all sides of us. There seems to me too much misery in the world. I cannot persuade myself that a beneficent and omnipotent God would have designedly created the Ichneumonidae with the express intention of their feeding within the living bodies of Caterpillars, or that a cat should play with mice. Not believing this, I see no necessity in the belief that the eye was expressly designed. On the other hand, I cannot anyhow be contented to view this wonderful universe, and especially the nature of man, and to conclude that everything is the result of brute force. I am inclined to look at everything as resulting from designed laws, with the details, whether good or bad, left to the working out of what we may call chance. Not that this notion at all satisfies me. I feel that the whole subject is too profound for the human intellect. A dog might as well speculate on the mind of Newton. Let each man hope and believe what he can. Certainly I agree with you that my views are not at all necessarily atheistical." From Francis Darwin, ed., *The Life and Letters of Charles Darwin* (London: John Murray, 1887), quoted in M. Mandelbaum, "Darwin's Religious Views," *Journal of the History of Ideas* 19, no. 3 (1958): 363, doi:10.2307/2708041.

215 **"raspberries and scent-free mints"**: J. J. Ewel, J. Mascaro, C. Kueffer, A. E. Lugo, L. Lach, and M. R. Gardener, "Islands: Where Novelty is the Norm," 29–44, in Richard J. Hobbs, Eric S. Higgs, and Carol Hall, eds., *Novel Ecosystems: Intervening in the New Ecological World Order* (Oxford: John Wiley and Sons, 2013).

216 **too are thinning out**: Elizabeth Wandrag and Haldre Rogers, "Guam's forests are being slowly killed off—by a snake," *The Conversation*, August 31, 2017, https://the conversation.com/guams-forests-are-being-slowly-killed-off-by-a-snake-83224.

216 **species are from islands**: Bernie R. Tershy, Kuo-Wei Shen, Kelly M. Newton, Nick D. Holmes, and Donald A. Croll, "The Importance of Islands for the Protection of Biological and Linguistic Diversity," *BioScience* 65, no. 6 (June 2015): 592–97, doi:10.1093/biosci/biv031.

216 **then cultivated at Kew**: K. Dehnen-Schmutz and M. Williamson, "Rhododendron ponticum in Britain and Ireland: Social, Economic and Ecological Factors in its Successful Invasion," *Environment and History* 12, no. 3 (2006): 325–50, doi:10.3197 /096734006778226355.

216 **wild in Cambridgeshire**: Robert A. Francis, ed., *A Handbook of Global Freshwater Invasive Species* (Abingdon, VA: Earthscan, 2012), 59.

216 **on a number 27 bus**: Pat Heslop-Harrison, "Great Escapes—continued," *Botany One*, July 26, 2010, https://www.botany.one/2010/07/great-escapes-continued/.

216 **escaped from botanical gardens**: P. E. Hulme, "Addressing the threat to biodiversity from botanic gardens," *Trends in Ecology and Evolution* 26, no. 4 (2011): 168–74, doi:10.1016/j.tree.2011.01.005.

Notes

217 "bubbly mass of red weed": H. G. Wells, *The War of the Worlds* (New York and London: Harper and Brothers, 1898; Fairbanks, AK: Project Gutenberg, 1992), ebook loc. 335.

217 a single source: Amani, Tanzania: Andy Coghlan, "Botanic gardens blamed for spreading plant invaders," *New Scientist* 2804, March 19, 2011.

221 forest in the region: Pierre Binggeli, "The ecology of Maesopsis invasion and dynamics of the evergreen forest of the East Usambaras, and their implications for forest conservation and forestry practices," in A. C. Hamilton and R. Bensted-Smith, eds., *Forest Conservation in the East Usambara Mountains, Tanzania* (Gland, Switzerland: IUCN, 1989), 269–300.

221 "or even local control": J. M. Hall, T. W. Gillespie, and M. Mwangoka, "Comparison of agroforests and protected forests in the East Usambara Mountains, Tanzania," *Environmental Management* 48 (2011): 237–47, cited in Hulme et al., "Aliens in the Arc."

221 and reclaiming their land: C. J. Kilawe, I. H. Mtwaenzi, and B. A. Mwendwa, "Maesopsis eminii Engl. mortality in relation to tree size and the density of indigenous tree species at the Amani Nature Reserve, Tanzania," *African Journal of Ecology* 56, no. 4 (2018): 1017–20, doi:10.1111/aje.12538; J. Viisteensaari, S. Johansson, V. Kaarakka, and O. Luukkanen, "Is the alien tree species *Maesopsis eminii* Engl. (Rhamnaceae) a threat to tropical forest conservation in the East Usambaras, Tanzania?" *Environmental Conservation* 27, no. 1 (2000): 76–81, doi:10.1017/s0376892900000096; also B. F. Nero and M. M. B. Mohamed, "Seedling dynamics under maesopsis tree canopy in different forest conditions at Amani Nature Reserve (ANR)," *Tropical Biology Association. Field course report at Amani Nature Reserve, Tanzania* (2005): 32–44.

222 "what has been, comes not again": Johann Wolfgang von Goethe, 1783, quoted in S. T. Jackson and R. J. Hobbs, "Ecological restoration in the light of ecological history," *Science* 325, no. 5940 (2009): 567–69.

223 the African tulip tree: Fred Pearce, "The Strange Case of the Puerto Rican Frog," *Anthropocene Magazine*, October 2016.

223 their new shared home: Pearce, "Strange Case."

223 damaged them in the first place: e.g., Carolina Murcia, James Aronson, Gustavo H. Kattan, David Moreno-Mateos, Kingsley Dixon, and Daniel Simberloff, "A critique of the 'novel ecosystem' concept," *Trends in Ecology & Evolution* (2014), doi: 10.1016/j.tree.2014.07.006.

224 covered by novel ecosystems: M. Perring and E. C. Ellis, "The extent of novel ecosystems: long in time and broad in space," in Richard J. Hobbs, Eric S. Higgs, and Carol M. Hall, *Novel Ecosystems: Intervening in the New Ecological World Order* (Hoboken, NJ: Wiley-Blackwell, 2013), 66–80.

224 the native species beneath: Heinke Jäger, "The rise and fall of the invasive quinine tree in Galápagos," *Galápagos Conservancy*, April 25, 2018, https://www.galapagos.org/blog/invasive-quinine-2018/.

225 "last vestiges out to sea": Wells, *The War of the Worlds*, ebook loc. 307.

Notes

Chapter 10: The Trip to Rose Cottage

232 document the island: I highly recommend Findlay's two books on the island, which present his photographs, archive images, and recollections of life on Swona by those who spent time there: John S. Findlay, *Swona: A Photographic Portrait* (Kirkwall, Scotland: Galaha Press, 2010); and John S. Findlay, *Swona Revisited* (Kirkwall, Scotland: Galaha Press, 2014). Ownership of the island is now held by the Annal family, descendants of Eva Rosie, who are in the process of creating a foundation for the preservation of the island's built and natural heritage. See www.swona.net for details.

233 out, Harold Wilson in: "Premier Wilson—I start now," *Press and Journal* (Aberdeen) March 5, 1974.

243 "been 'banished' there": S. J. G. Hall and G. F. Moore, "Feral cattle of Swona, Orkney Islands," *Mammal Review* 16, no. 2 (1986): 89–96, doi:10.1111/j.1365-2907:1986 .tb00026.x.

243 competition between the males: Findlay, *Swona Revisited*, 187–88.

244 "cattle as reverential": Findlay, *Swona Revisited*, 200.

246 in the African savannah: Sandy Annal, quoted in Findlay, *Swona Revisited*, 200.

247 "some country drooping ears": Charles Darwin, *On the Origin of Species: 150th Anniversary Edition* (1859; New York: Signet Classics, 2009), 34.

247 repeated with their offspring: A good, concise account of the Russian fox experiments can be found in Jason G. Goldman, "Man's new best friend? A forgotten Russian experiment in fox domestication," *Scientific American* guest blog, September 6, 2010, https:// blogs.scientificamerican.com/guest-blog/mans-new-best-friend-a-forgotten-russian -experiment-in-fox-domestication/.

250 over ten thousand years ago: Ruth Bollongino, Joachim Burger, et al., "Modern Taurine Cattle Descended from Small Number of Near-Eastern Founders," *Molecular Biology and Evolution* 29, no. 9 (September 1, 2012), 2101–4, https://doi.org/10 .1093/molbev/mss092.

250 "not even when young": *Sketches in Natural History: History of the Mammalia*, vol. 4 (London: C. Cox, 1849), 141. Caesar called the animal in question the "uri," although it is believed that he is discussing what we now know of as the aurochs. Translations vary; the W. A. McDevitte and W. S. Bohn translation of 1870 (New York: Harper and Brothers) offers: "these are a little below the elephant in size, and of the appearance, color, and shape of a bull. Their strength and speed are extraordinary; they spare neither man nor wild beast which they have espied . . . not even when taken very young can they be rendered familiar to men and tamed."

251 Nazi theories of eugenics: C. Driessen and J. Lorimer, "Back-breeding the aurochs: the Heck brothers, National Socialism and imagined geographies for nonhuman Lebensraum," in P. Giaccaria and C. Minca, *Hitler's Geographies* (Chicago: University of Chicago Press, 2016), 138–57.

251 "in the Third Reich": Driessen and Lorimer, "Back-breeding the aurochs," 138–57.

252 Frankenstein's monsters run amok: Tom Bawden, "Nazi super cows: British farmer forced to destroy half his murderous herd of bio-engineered Heck cows after they try to kill staff," *The Independent*, January 5, 2015.

253 at sites like Knepp, in England: An interesting discussion of the role of grazing animals in creating complex habitats can be found in Isabella Tree's inspiring account of their rewilding of the Knepp estate, *Wilding: The Return of Nature to a British Farm* (London: Picador, 2018).

253 essay, "Faking Nature": Robert Elliot, "Faking nature," *Inquiry: An Interdisciplinary Journal of Philosophy* 25, no. 1 (1982): 81–93.

254 new kingdom of the cattle: Findlay, *Swona Revisited*, 187: "Fifty years—say ten cattle generations" as of 2015.

254 require assistance in birth: Jennifer Bentley, "Extension and Outreach: Calving Process and Assistance," Iowa State University, March 2016.

255 bull to an Angus cow: Hall and Moore, "Feral cattle of Swona, Orkney Islands."

255 generations and a thousand: Michael M. Desai, "Reverse evolution and evolutionary memory," *Nature Genetics* 41 (2009): 142–43, doi:10.1038/ng0209-142.

255 abandonment and isolation: "In brief: Unique herd found on island," *The Guardian*, October 21, 1999.

256 less thoroughly domesticated: C. G. Thulin, Paulo C. Alves, et al., "Wild opportunities with dedomestication genetics of rabbits," *Restoration Ecology*, February 2017, https://doi.org/10.1111/rec.12510.

256 shot dead by conservationists: R. Rozzi and M. V. Lomolino, "Rapid Dwarfing of an Insular Mammal—The Feral Cattle of Amsterdam Island," *Scientific Reports* 7, no. 1 (2017), 8820, doi:10.1038/s41598-017-08820-2.

PART FOUR: ENDGAME

Chapter 11: Revelation

266 to burn the nostrils: Sharmen Greenaway, *Montserrat in England: Dynamics of Culture* (Bloomington, IN: iUniverse, 2011), 10.

266 machinery grinding into action: Greenaway, *Montserrat in England*, 11.

267 Get out of there: David Lea's self-published memoir, which records the eruptions in detail and with fascinating local insight, is published as *Through My Lens* (2015) and it, along with Lea's documentaries, *The Price of Paradise*, are available at priceofparadise.com.

269 vitrify your brain: Brigit Katz, "Vesuvius' Scorching Eruption Turned a Man's Brain Into Glass," Smithsonianmag.org, January 23, 2020, https://smithsonianmag.com/smart-news/vesuvius-scorching-eruption-turned-mans-brain-glass-180974041/.

Notes

269 **flesh from your bones:** P. Petrone, P. Pucci, A. Vergara, A. Amoresano, L. Birolo, et al., "A hypothesis of sudden body fluid vaporization in the 79 AD victims of Vesuvius," *PLOS One* 13, no. 9 (2018): e0203210, doi:10.1371/journal.pone.0203210.

269 **somewhere out of sight:** A full study of the (earlier, most devastating phases of) volcanic activity on Montserrat can be found in Timothy H. Druitt and B. Peter Kokelaar, *The Eruption of Soufrière Hills Volcano, Montserrat, from 1995 to 1999* (London: Geological Society of London, 2002). The Soufrière Hills volcano continued to erupt until 2010, and is not yet officially "dormant"; research led by Professor Locko Neuberg at the Center for the Observation and Modelling of Earthquakes, Volcanoes and Tectonics at Leeds University reported in 2016 that magma continues to collect in a reservoir under the island and eruption is therefore "far from over." For more details, see https://comet.nerc.ac.uk/montserrat-continues -inflate/.

270 **otherwise escaped unharmed:** Lea, *Through My Lens*, 123; also discussed in S. C. Loughlin, P. J. Baxter, W. P. Aspinall, B. Darroux, C. L. Harford, and A. D. Miller, "Eyewitness accounts of the 25 June 1997 pyroclastic flows and surges at Soufrière Hills Volcano, Montserrat, and implications for disaster mitigation," *Geological Society, London, Memoirs* 21, no. 1 (2002): 211–30, doi:10.1144/GSL.MEM.2002.021.01.10.

270 **faceful of scorching ash:** Loughlin et al., "Eyewitness accounts."

271 **"would never believe it":** Loughlin et al., "Eyewitness accounts."

271 **by a wall of fire:** Collected firsthand accounts of the Mount St. Helens eruption in Dana Hunter, "The Cataclysm: Vancouver! Vancouver! This Is It!" *Scientific American: Rosetta Stones,* August 9, 2012, https://blogs.scientificamerican.com/rosetta-stones /the-cataclysm-vancouver-vancouver-this-is-it/.

271 **are therefore blasphemous:** D. K. Chester, "The Theodicy of Natural Disasters," *Scottish Journal of Theology* 51, no. 4 (1998): 485, doi:10.1017/s0036930600056866.

274 **appear around the sun:** Now known as "Bishop's rings" after the correspondent, Rev. S. A. Bishop; see Royal Society (Great Britain) Krakatoa Committee, *The Eruption of Krakatoa, and Subsequent Phenomena* (London: Royal Society, 1888), 262–63.

274 **"an immense crimson curtain:"** Royal Society, *Eruption of Krakatoa,* 173.

274 **"high in the heavens":** *Hanover Spectator,* December 19, 1883, quoted in Donald W. Olson, Russell L. Doescher, and Marilynn S. Olson, "When the Sky Ran Red: The Story Behind *The Scream,*" *Sky & Telescope,* February 2004.

274 **blast in recorded history:** William J. Broad, "A Volcanic Eruption That Reverberates 200 Years Later," *The New York Times,* August 24, 2015.

275 **forecasts of mass destruction:** Gillen D'Arcy-Woods, *Tambora: The Eruption that Changed the World* (Princeton, NJ: Princeton University Press, 2015), offers an excellent overview of the global chaos that broke out in the wake of the Tambora eruption.

275 **as far south as Virginia:** Robert Evans, "Blast from the Past," *Smithsonian Magazine,* July 2002.

Notes

275 **of four or five years:** M. Stoffel, Christophe Corona, et al., "Estimates of volcanic-induced cooling in the Northern Hemisphere over the past 1,500 years," *Nature Geoscience* 8, no. 10 (2015): 784–88, cited in C. M. Vidal, N. Métrich, J.-C. Komorowski, and I. Pratomo, "The 1257 Samalas eruption (Lombok, Indonesia): the single greatest stratospheric gas release of the Common Era," *Scientific Reports* 6, no. 34868 (2016), https://doi.org/10.1038/srep34868.

275 **"in the muddy street":** Quoted in C. Newhall, S. Self, and A. Robock, "Anticipating future Volcanic Explosivity Index (VEI) 7 eruptions and their chilling impacts," *Geosphere* 14, no. 2 (2018): 572–603, doi:10.1130/ges01513:1.

276 **thousand years, or thereabouts:** Newhall et al., "Anticipating future Volcanic Explosivity Index."

276 **once in every 30,000:** Sebastian Farquhar, John Halstead, Owen Cotton-Barratt, Stefan Schubert, Haydn Belfield, and Andrew Snyder-Beattie, *Existential Risks: Diplomacy and Governance* (Oxford: Global Priorities Project, 2017), 10.

276 **rose by around 18°F:** Becky Oskin, "Earth's Greatest Killer Finally Caught," *Live Science*, December 12, 2013, https://www.livescience.com/41909-new-clues-permian-mass-extinction.html.

277 **speaking, is due to do:** Brian Wilcox of NASA's Jet Propulsion Laboratory, in 2017: "Yellowstone explodes roughly every 600,000 years, and it is about 600,000 years since it last exploded," quoted in David Cox, "Nasa's ambitious plan to save Earth from a supervolcano," *BBC Future*, August 17, 2017.

277 **for a decade or more:** C. Timmreck and H.-F. Graf, "The initial dispersal and radiative forcing of a northern hemisphere mid-altitude supervolcano: a model study," *Atmospheric Chemistry and Physics* 6, no. 1 (2006); also cited in Bryan Walsh, *End Times: A Brief Guide to the End of the World* (New York: Hachette, 2019).

279 **moments of their demise:** See the excellent narrative of Pompeii's last moments in Dana Hunter, "How Pompeii Perished," "Rosetta Stones" blog, *Scientific American*, November 27, 2012.

283 **an estimated 1.8°F:** J. Kandlbauer, P. O. Hopcroft, P. J. Valdes, and R. S. J. Sparks, "Climate and carbon cycle response to the 1815 Tambora volcanic eruption," *Journal of Geophysical Research: Atmospheres* 118, no. 22 (2013): 12497–507, doi:10.1002/2013jd019767.

284 **by 2100 is likely:** An excellent statistical analysis of emissions and future temperature trends can be found in J. Tollefson, "Can the world kick its fossil-fuel addiction fast enough?" *Nature* 556, no. 7702 (2018): 422–25, doi:10.1038/d41586-018-04931-6.

284 **"just a couple of centuries":** Quoted in Melissa Davey, "Humans causing climate to change 170 times faster than natural forces," *The Guardian*, February 12, 2017.

284 **"than a gradual change":** O. Gaffney and W. Steffen, "The Anthropocene equation," *The Anthropocene Review* 4, no. 1 (2017): 53–61, doi:10.1177/2053019616688022.

285 **"upon everlasting generations?":** Mary Shelley, *The Annotated Frankenstein*, ed. Susan J. Wolfson and Ronald L. Levao (Cambridge, MA: Harvard University Press, 2012), 255.

Notes

285 **"a new biological world order":** Chris D. Thomas, *Inheritors of the Earth* (London: Allen Lane, 2017), 91–92.

288 **residence in an abandoned house:** Steve H. Holliday, *Montserrat: A Guide to the Center Hills* (St. John's, Antigua: West Indies Publishing Ltd., 2009), 128.

Chapter 12: The Deluge and the Desert

293 **much as six inches a day:** "Salton Sea Rises Daily," *The Los Angeles Herald* 33, no. 259, June 16, 1906.

293 **and then their homes:** Marc Reisner, *Cadillac Desert: The American West and Its Disappearing Water* (New York: Penguin Books, 1986), 123.

294 **"to be found anywhere":** "The Salton Sea is here to stay," *The Hanford Sentinel* 20, no. 32, August 17, 1905.

296 **"at the end, poison":** Sven Erik Jorgenson, ed., *Encyclopedia of Environmental Management*, vol. 1 (Boca Raton, FL: CRC Press, 2013), 302.

296 **at 5 percent salinity:** M. A. Tiffany, J. Wolny, M. Garrett, K. Steidinger, and S. H. Hurlbert, "Dramatic blooms of *Prymnesium* sp. (Prymnesiophyceae) and *Alexandrium margalefii* (Dinophyceae) in the Salton Sea, California," *Lake and Reservoir Management* 23, no. 5 (2007): 620–29, doi:10.1080/07438140709354041.

296 **of fish in an hour:** J. P. Cohn, "Saving the Salton Sea; researchers work to understand its problems and provide possible solutions," *BioScience* 50, no. 4 (2000): 295–301.

297 **the size of satsumas:** Becky Oskin, "Rotting Balls of Fish Flesh Invade Salton Sea's Shores," *Live Science*, October 30, 2013, https://www.livescience.com/40809-salton-sea-dead-fish-balls.html.

297 **more than 200,000 grebes:** W. W. Carmichael and R. Li, "Cyanobacteria toxins in the Salton Sea," *Aquatic Biosystems* 2, no. 5 (2006), doi:10.1186/1746-1448-2-5.

299 **"solution from the sea":** Peter Ward, *The Medea Hypothesis: Is Life on Earth Ultimately Self-Destructive?* (Princeton, NJ: Princeton University Press, 2015), ebook loc. 84

299 **"certainly could happen again":** Ward, *The Medea Hypothesis*, 71.

300 **"selfish and ultimately biocidal":** Ward, *The Medea Hypothesis*, xx.

300 **1,000 parts per million (ppm):** Ward, *The Medea Hypothesis*, 82.

300 **horrifying 2,000 ppm by 2250:** Nicola Jones, "How the World Passed a Carbon Threshold and Why It Matters," *Yale Environment 360*, Yale School of Forestry & Environmental Studies, January 26, 2017, https://e360.yale.edu/features/how-the-world-passed-a-carbon-threshold-400ppm-and-why-it-matters.

303 **"level sands stretch far away":** Zachary Leader and Michael O'Neill, eds., *Percy Bysshe Shelley: The Major Works* (Oxford World Classics) (2003; Oxford: Oxford University Press, 2009), 198.

Notes

312 **"and war" feature heavily:** Jem Bendell, "Deep Adaptation: A Map for Navigating Climate Tragedy," an occasional paper from the Institute of Leadership and Sustainability, July 27, 2018, http://lifeworth.com/deepadaptation.pdf.

312 **countries within a decade:** From Jem Bendell's contribution to the Extinction Rebellion handbook, *This is Not a Drill* (London: Penguin Books, 2019); a longer version is available online at https://jembendell.com/2020/01/15/adapting-deeply-to-likely-collapse-an-enhanced-agenda-for-climate-activists/.

312 **"collapse or catastrophe begin?":** Bendell, "Deep Adaptation."

313 **next ten to fifteen years:** Jennifer Harper, "51% of young voters believe life on Earth will end in the next 10–15 years: Poll," *The Washington Times*, September 24, 2019.

313 **nature is not far away:** P. R. Ehrlich and A. H. Ehrlich, "The Population Bomb Revisited," *The Electronic Journal of Sustainable Development* 1, no. 3 (2009). See also Damian Carrington, "Paul Ehrlich: 'Collapse of civilization is a near certainty within decades,'" *The Guardian*, March 22, 2018.

314 **appeal of disanthropic thinking:** See Greg Garrard on "disanthropy," the yearning for the absence or negation of humans, which he contrasts with the more common "misanthropy," the dislike of all mankind, in R. Ghosh and G. Garrard, "Worlds Without Us: Some Types of Disanthropy," *SubStance* 41, no. 1 (2012): 40–60, doi:10.1353/sub.2012.0001.

315 **"nature is to kill ourselves":** William Cronon, "The Trouble With Wilderness; Or, Getting Back to the Wrong Nature," in William Cronon, ed., *Uncommon Ground: Rethinking the Human Place in Nature* (New York: W. W. Norton & Co., 1995), 69–90.

315 **"pass away—time it did":** D. H. Lawrence, *Women in Love* (1920; New York: Barnes & Noble Classics, 2005), ebook loc. 165.

319 *"utterly burned with fire":* Revelation 18:8.

319 *"hour is thy judgment come":* Revelation 18:10.

320 *"in the land of great drought":* Hosea 13:5.

323 **"overtreatment and therapeutic nihilism":** Louis Lasagna, "Hippocratic oath," modern version, 1964.

324 **"and all manner of thing shall be well":** Recorded by Julian, Anchoress at Norwich, in 1373, and published as *Revelations of Divine Love* (London: Methuen & Co., 1901; Fairbanks, AK: Gutenberg Project, 2016), ebook loc. 1681.

325 **"a death, rebirth cycle":** Melody Sample, quoted in Rory Carroll, "In a forgotten town by the Salton Sea, newcomers build a bohemian dream," *The Guardian*, April 23, 2018.

326 **numbered over a million:** Stephanie Weagley and Carol A. Roberts, Carlsbad Fish and Wildlife Office, "Field notes: Desert Pupfish and the Salton Sea's Experimental Research Ponds," U.S. Fish and Wildlife Service (California-Nevada), August 31, 2010.

INDEX

Index

Index

Index

Index

Index

Index

Index

Index

West Kirby (near Liverpool), 176
wetland ecosystems, 31, 33, 53, 59
whales, 167
White, Lynn, Jr., 34
Whiteread, Rachel, *House*, 281
Williams, William Carlos, 151, 156
    Paterson, 137, 142
Wilson, E. O., 323
wolves, 75, 96–97, 101
World Health Organization, 102

Xenophon, 327

Yellowstone National Park, 276, 276
Yorke Bay (Falkland Islands), 56
Young's helleborine (orchid), 20
Yucatán civilization, 157

Zambia, Copperbelt of, 189
Zimmerman, Albrecht, 203